Flow Measurement Engineering Handbook

Flow Measurement Engineering Handbook

Editor

Manish Bhardwaj

scitus
academics

Flow Measurement Engineering Handbook
Edited by **Manish Bhardwaj**

Printed in 2017

ISBN: 978-1-68117-372-6

Library of Congress Control Number: 2015941560

© 2016 by

SCITUS Academics LLC,
616, Corporate Way, Suite 2, 4766,
Valley Cottage, NY 10989

www.scitusacademics.com

Contents

Preface vii

Chapter 1   **CFD Simulation of the Discharge Flow from Standard Rushton Impeller** 1

Bohuš Kysela, Jiří Konfršt, Ivan Fořt, and Zdeněk Chára

Chapter 2   **Engineering Parameters in Bioreactor's Design: A Critical Aspect in Tissue Engineering** 19

Nasim Salehi-Nik, Ghassem Amoabediny, Behdad Pouran, Hadi Tabesh, Mohammad Ali Shokrgozar, Nooshin Haghighipour, Nahid Khatibi, Fatemeh Anisi, Khosrow Mottaghy, and Behrouz Zandieh-Doulabi

Chapter 3   **CFD Prediction of the Turbulent Flow Generated in Stirred Square Tank by a Rushton Turbine** 67

W. Chtourou, M. Ammar, Z. Driss, and M. S. Abid

Chapter 4   **Discrete Tracer Point Method to Evaluate Turbulent Diffusion in Circular Pipe Flow** 95

Arif Widiatmojo, Kyuro Sasaki, Nuhindro Priagung Widodo, and Yuichi Sugai

Chapter 5   **Electrical Capacitance Probe Characterization in Vertical Annular Two-Phase Flow** 123

Grazia Monni, Mario De Salve, Bruno Panella, and Carlo Randaccio

Chapter 6   **Characterization of two Phase Flows in Chemical Engineering Reactors** 157

S.L. Kiambi, H.K. Kiriamiti, and A. Kumar

Chapter 7   **An Integrative Image Measurement Technique for Dense Bubbly Flows with a Wide Size Distribution** 183

Ashish Karn, Christopher Ellis, Roger Arndt, and Jiarong Hong

Chapter 8   **Multiphase Monolith Reactors: Chemical Reaction Engineering of Segmented Flow in Microchannels** 215

Michiel T. Kreutzer, Freek Kapteijn, Jacob A. Moulijn, Jand ohan J. Heiszwolf

Citations... 281
Index.. 285

Preface

Flow measurement is the quantification of bulk fluid movement. Flow can be measured in a variety of ways. Positive-displacement flow meters accumulate a fixed volume of fluid and then count the number of times the volume is filled to measure flow. Other flow measurement methods rely on forces produced by the flowing stream as it overcomes a known constriction, to indirectly calculate flow. Flow may be measured by measuring the velocity of fluid over a known area. As noted in the preceding Dedication, the tendency to make flow measurement a highly theoretical and technical subject overlooks a basic tenet: Practical application of meters, metering principles, and metering instrumentation and related equipment is the real key to quality measurement. And that includes the regular maintenance by trained and experienced personnel with quality equipment required to keep flow measurement systems operating so as to achieve their full measurement potential.

Editor

Chapter 1

CFD Simulation of the Discharge Flow from Standard Rushton Impeller

Bohuš Kysela[1], Jiří Konfršt[1], Ivan Fořt[2], and Zdeněk Chára[1]

[1]Institute of Hydrodynamics AS CR, v. v. i., Pod Patankou 30/5, 166 12 Prague, Czech Republic

[2]Department of Process Engineering, Faculty of Mechanical Engineering, Czech Technical University in Prague, Technicka 4, 166 07 Prague, Czech Republic

ABSTRACT

The radial discharge jet from the standard Rushton turbine was investigated by the CFD calculations and compared with results from the Laser Doppler Anemometry (LDA) measurements. The Large Eddy Simulation (LES) approach was employed with Sliding Mesh (SM) model of the impeller motion. The obtained velocity profiles of the mean ensemble-averaged velocity and r.m.s. values of the fluctuating velocity were compared in several distances from the impeller blades. The calculated values of mean ensemble-averaged velocities are rather in good agreement with the measured ones as well as the derived power number from calculations. However, the values of fluctuating velocities are obviously lower from LES calculations than from LDA measurements.

INTRODUCTION

The flow inside the agitated vessel has a key role in the mixing processes. Only the CFD modeling gives us the complex information about the whole flow field in contrast with the results of the experimental measurements. The enormous progress of the computational equipment has allowed using more exacting turbulence models for solution of the flow in the agitated vessel. Nowadays not only Reynolds Averaged Navier-Stokes (RANS) models (for example, k-ε [1–7], k-ω, and Reynolds Stress Model [4, 5]) are commonly used, but also other more sophisticated methods become topical (for example, Detached Eddy Simulation [8], Large Eddy Simulation [8–15] and even Direct Numerical Simulation [15]). However, all calculations also need some validation by the experimental results or by the analytical models. The radial impellers are most often used in experiments and calculations [16–21], namely, Rushton turbine, and there are also analytical models, where the impeller discharge stream is modeled as a turbulent jet [1, 2, 16, 19, 20].

The aim of this study is the description of the turbulent velocity field in the discharge stream from the standard Rushton turbine impeller in the pilot plant mixing vessel with baffles at the wall. Investigation will be carried out experimentally (LDA technique) as well as by means of CFD simulation, where the LES approach was used with Sliding Mesh model for the impeller movement because Sliding Mesh approach has a potential to be used as a design tool to screen different configurations, but it has not been sufficiently validated for turbulent regime [4].

CFD CALCULATIONS

A commercial ANSYS FLUENT v.13.0 solver of the finite volume method was employed. The turbulence was modelled by Large Eddy Simulation (LES) with Sliding Mesh (SM) simulation for the impeller movement. The solver was pressure based and for pressure-velocity coupling the PISO method was used. The Smagorinsky-Lilly model was used as the subgrid-scale model with second-order implicit scheme. The schemes for spatial discretization were Gradient—Least Squares Cell based, Pressure—Second Order, and Momentum—Second Order Upwind. The boundary conditions were set: water level to the symmetry and others to the no slip wall, where the part of the impeller shaft outside of the sliding region is defined as wall with impeller speed velocity. The walls of the vessel and baffles are provided by the boundary layer mesh; see Figure 1. The sliding region has cylindrical shape with distance d/10 from the impeller cylindrical envelope; see Figure 2. The solved hexahedral meshes consists of 2 465 228 cells (LES 1) and 7 435 557 cells (LES 2), respectively. The larger mesh was refined, namely, in bulk vessel region to attained the maximal cell size under 2 mm, which corresponds with maximal size of the measurement LDA volume; see Section 3. The time step must not exceed 1/60 of one revolution [11] that corresponds with 0.0032 s for 300 rpm. Hence, the time step 0.001 s was used. The calculated time was 60 s while the flow development required min 20 revolutions [8, 13] which represent the first 2 s of the simulation. Calculated instantaneous flow field over 60 s in the measured plane is depicted in Figure 3.

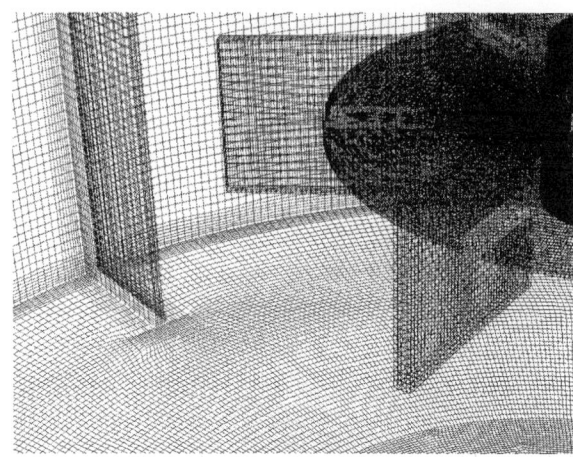

Figure 1: Mesh in the baffles vicinity and on the impeller surface in detail (LES 1).

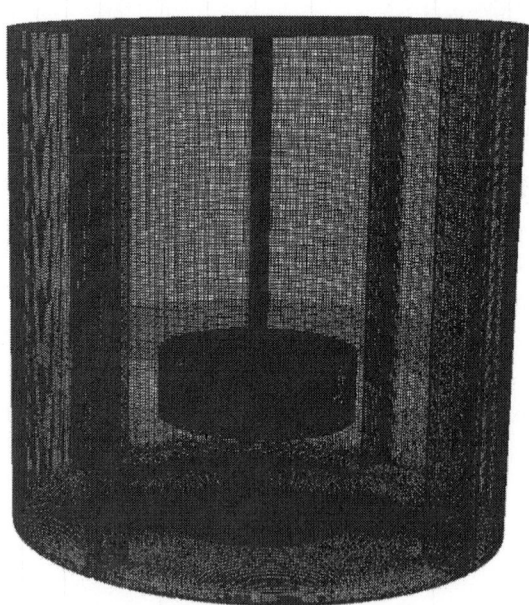

Figure 2: Mesh (LES 1) on the vessel wall and impeller with the high-lighted sliding region around the impeller.

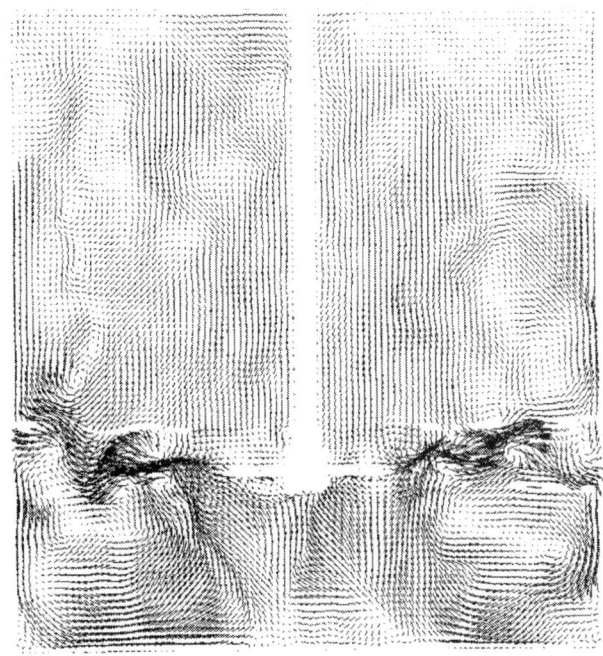

Figure 3: Calculated instantaneous flow field in measured plane between the baffles.

EXPERIMENTAL

Measurements of the velocity profiles were carried out in a pilot plant flat bottomed mixing vessel with four baffles at its wall (see Figure 4), with water as the working liquid (density p=1000 kg m^{-3}, dynamic viscosity μ-1 mPa s) under the constant impeller speed 300 rpm (impeller Reynolds number Re$_M$=50000). A standard Rushton impeller [22, 23] was used for the investigation (see Figure 5).

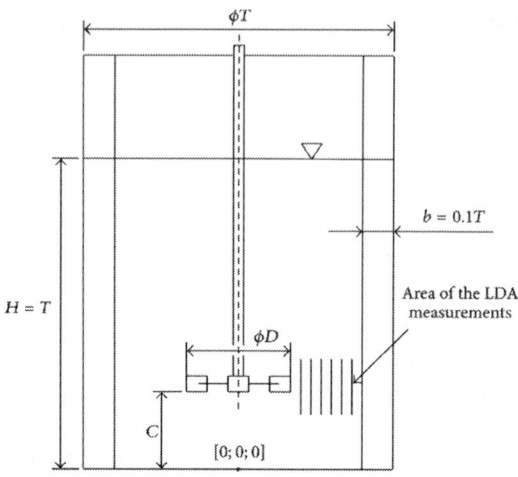

Figure 4: Pilot plant cylindrical vessel with a six-blade Rushton turbine (T=300 mm, H/T=1,C/D=0.75 b/T=1/10, four baffles).

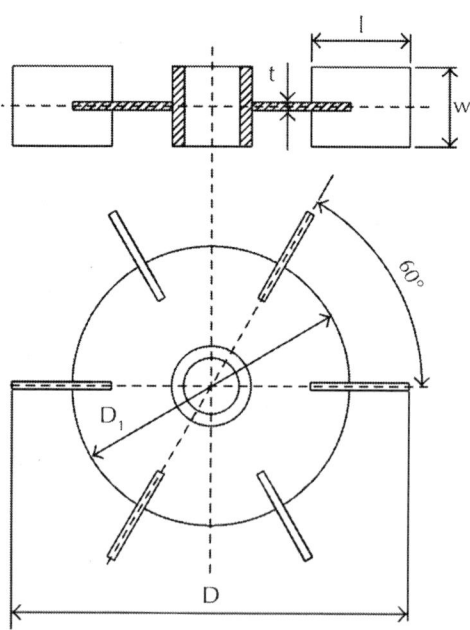

Figure 5: Sketch of Rushton turbine impeller, $\omega/D=0.2$, $D_1/D=0.75$, and l/D=0.25.

Laser Doppler Anemometry (LDA) one-component measurements of the radial velocity were performed in the impeller discharge stream (see Figure 4) in the vertical plane between two adjacent baffles in axial profiles with several distances from the impeller blade. The dimensionless radial coordinates $r^*=2r/D$ were 1.2; 1.4; 1.6; 1.8; 2.0; 2.2. One-component LDA system setup consists of Coherent INNOVA 305 Ion-Argon laser supply with power 5 W and separated beam only for one component measurement on wavelength 514.5 nm; DANTEC FiberFlow transmitting optics and P80 DANTEC BSA processor. The optic parameters were focal length 310 mm, diameter of the beam 1.2 mm, fringe spacing 2.67 μm, number of fringes 63, and the measured volume was ellipsoid with sizes of the axes 0.170×0.169×1.757 mm. The used frequency shift was 40 MHz and velocity span 7.51 m/s. The setup was supervised by BSA FLOW SOFTWARE v3.0 installed on standard PC where the data was processed. S-HGS (silver coated-hollow glass Spheres) with mean diameter 10 μm and density 1.1 gcm^{-3} were used as trace particles. The measurement was performed through the glass flat bottom of the vessel to eliminate optical effects of the cylindrical walls.

RESULTS AND DISCUSSION

Profiles of the radial mean ensemble-averaged velocity component in the dimensionless form where the radial velocity component is normalized by the impeller tip speed $V_{tip} = \pi Dn$ are depicted in Figures 6, 7, 8, 9, 10, and 11. The dimensionless coordinate z^* is the distance from the impeller disk axis normalized by the half-height of the impeller blade. The profiles are depicted for six values of the dimensionless radius r^*.

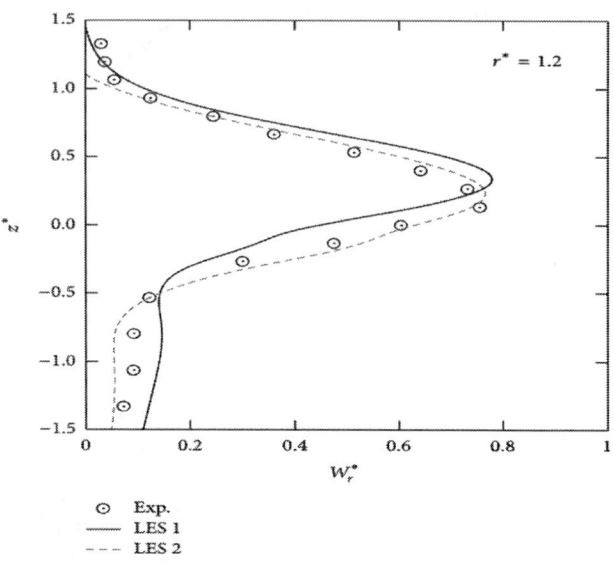

Figure 6: Radial component of the ensemble-averaged mean velocity at r*=1.2.

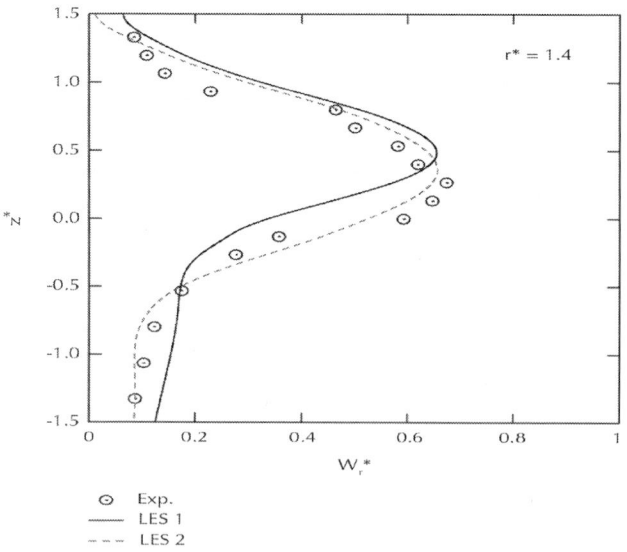

Figure 7: Radial component of the ensemble-averaged mean velocity at r*=1.4.

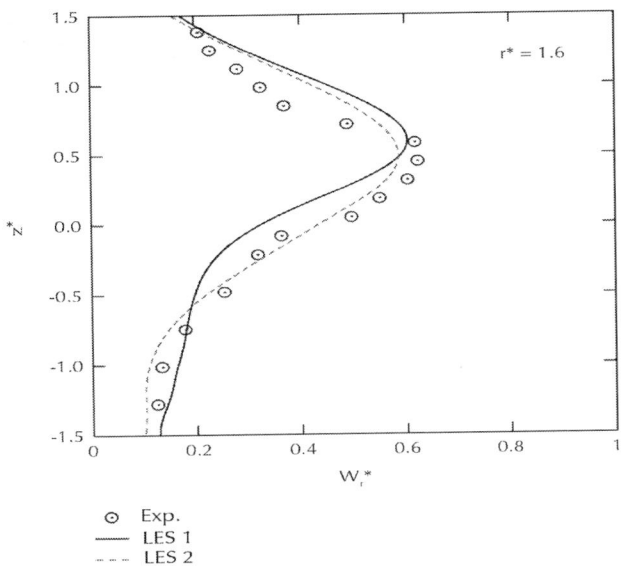

Figure 8: Radial component of the ensemble-averaged mean velocity at r*=1.6.

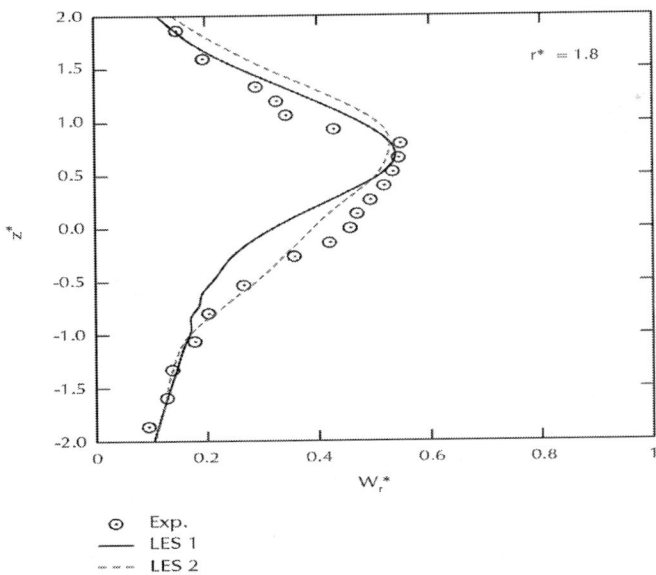

Figure 9: Radial component of the ensemble-averaged mean velocity at r*=1.8.

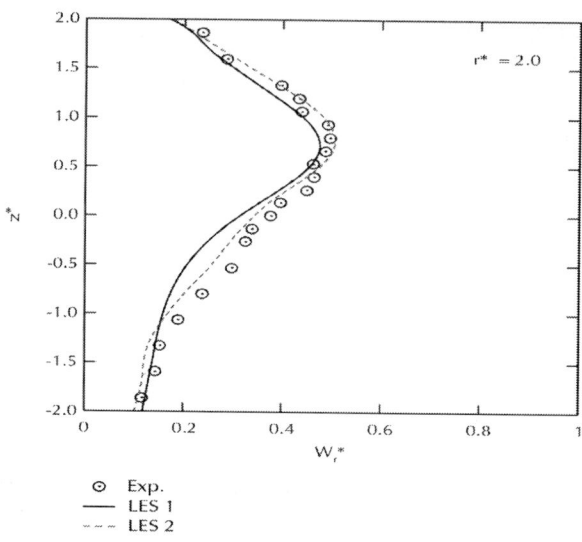

Figure 10: Radial component of the ensemble-averaged mean velocity at r*=2.0.

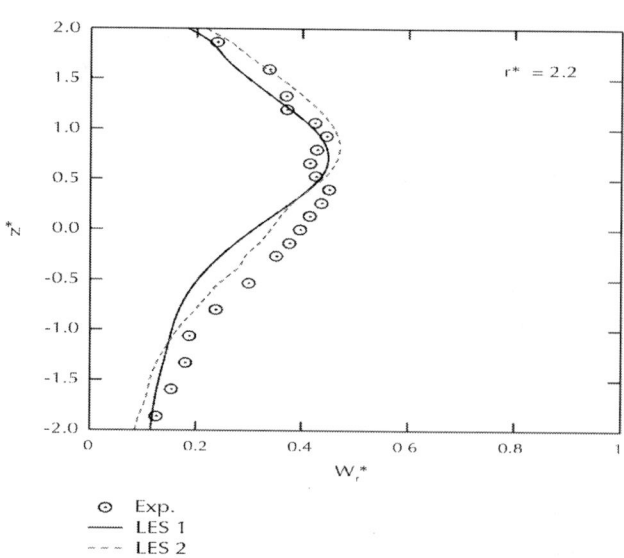

Figure 11: Radial component of the ensemble-averaged mean velocity at r*=2.2.

The depicted results from CFD calculations and results in measured points were compared also by calculations of variance. Variance between the measured and the calculated data of dimensionless radial component of the mean ensemble-averaged velocity was calculated using the formula

$$\mathrm{var}\left(W_r^*\right) = \frac{1}{N}\sum_{i=1}^{N}\left(W_{r\,\mathrm{LDA}}^* - W_{r\,\mathrm{CFD}}^*\right)^2,$$

(1)

where N=24 is the number of compared points on the profile W_r^* with index LDA means the value obtained from LDA measurements and with index CFD is the value interpolated from calculated profile from CFD. Standard deviation σ is expressed as the square root of the variance. The results of variance (standard deviation) in Table 1 signify that the higher discrepancy of the profiles is in the region where the zone of establishment is changing to the zone of the established flow [1, 2]. The increase of the standard deviation with increasing dimensionless radius is probably caused by the different shape of the discharge stream and it seems to be different also at the vertical position of the stream which depends on the impeller off-bottom clearance.

Table 1: Mean variance value between LDA data of the mean ensemble-averaged radial velocity component and profiles obtained from CFD in dimensionless form

r^*	LES 1		LES 2	
	Var (W_r^*)	σ	Var (W_r^*)	σ
1.2	0.0055	0.074	0.0017	0.042
1.4	0.0082	0.090	0.0026	0.051
1.6	0.0056	0.075	0.0024	0.049
1.8	0.0042	0.065	0.0032	0.057
2.0	0.0017	0.042	0.0006	0.024
2.2	0.0027	0.052	0.0017	0.041

The r.m.s. values of the fluctuation velocity were treated as well as the mean velocities to the dimensionless form. The two values of the dimensionless radius r*=1.4 and 1.6 are shown in Figure 12, where the trailing vortices have an impact on the fluctuation velocity. The zone is titled the zone of establishment (ZFE) [1]. The calculated values of the fluctuations are rather lower than the measured ones. The results in the next zone titled the zone of established flow (ZEF) are depicted in Figure 13. There are compared values of the dimensionless radius r*=2.0 and 2.2. It seems that agreement between computed and experimentally determined values of the fluctuating velocity is better in ZEF than in ZFE..

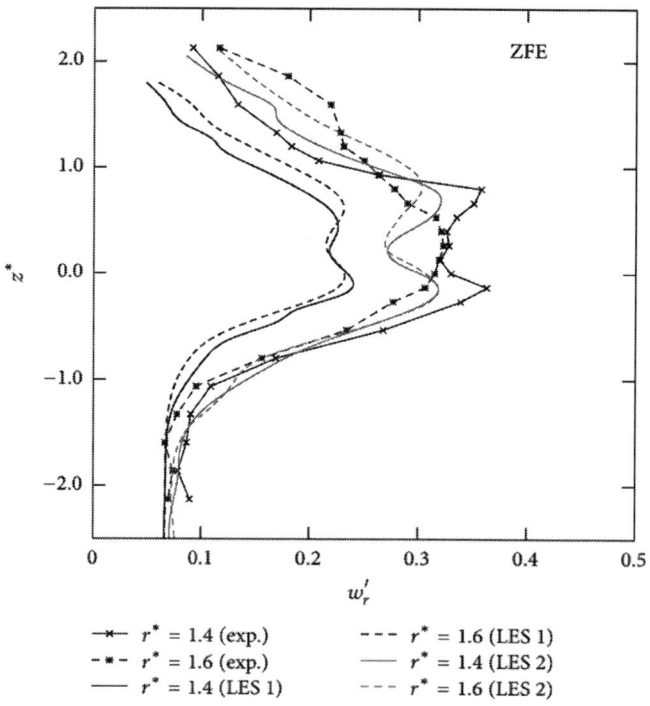

Figure 12: Comparison of axial profiles of r.m.s. values of radial fluctuation velocity in the zone of establishment (ZFE).

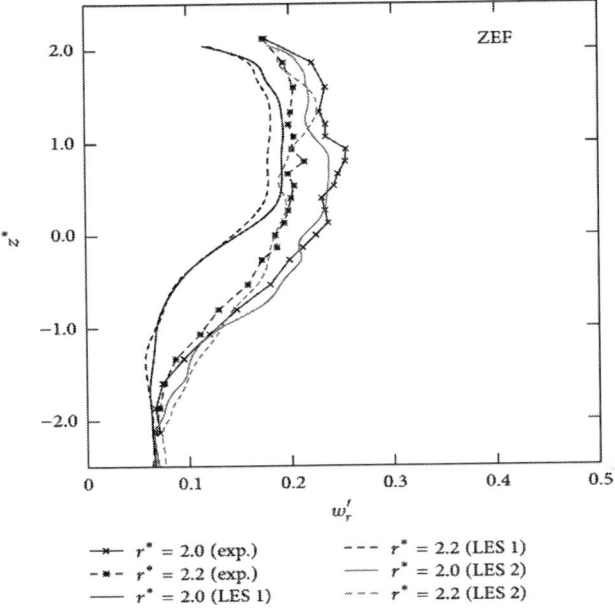

Figure 13: Comparison of axial profiles of r.m.s. values of radial fluctuation velocity in the zone of the established flow (ZEF).

The power number Po was calculated from impeller torque M_K which was obtained from the force balance on the impeller surface provided by the CFD calculations [8]:

$$P = 2\pi n M_k,$$

$$Po = \frac{P}{\rho n^3 D^5}. \tag{2}$$

For Rushton turbine, the results of the power number could be compared with the empirical correlation [22,23], relating the power number to the relative blade thickness t/D and the relative vessel diameter T/T_0:

$$Po = 2.512\left(\frac{t}{D}\right)^{-0.195}\left(\frac{T}{T_0}\right)^{0.063}, \tag{3}$$

where t=2 mm is the thickness of the separating disc of a standard Rushton impeller and quantity T_0=1m. The power number derived from calculations (2) was Po=5.32 and it is in a good agreement with power number calculated by (3), where Po=5.00.

CONCLUSIONS

The flow in the discharge stream from the standard Rushton turbine was calculated by the Large Eddy Simulation approach. The comparison of the mean radial ensemble-averaged velocity profiles obtained from LDA measurements are in good agreement with the calculated results from both LES cases. The r.m.s. values of fluctuating velocity show the similar shape of profiles, but the calculations mostly underestimated the values obtained by the LDA measurements. The ensemble-averaging shows the dependency on the spatial resolution of the calculations or the measurement method; namely, the r.m.s. values of fluctuations are strongly affected by a spatial averaging. Calculated values of standard deviation from the mean velocities are increasing from lower dimensionless radius to the value r*=1.8 where it probably indicates the boundary between the zone of the flow establishment and the zone of the established flow. The same trend is shown in the comparisons of the r.m.s. values of the fluctuating velocities. The power number Po=5.32 derived from the impeller torque calculations of the presented LES numerical modeling is very close to the value Po=5.00 estimated from empirical correlation based on experimental measurements.

ACKNOWLEDGMENTS

This research has been subsidized by the Research Project no. GA CR P101/12/2274 and RVO: 67985874.

REFERENCES

1. R. Ben-Nun and M. Sheintuch, "Characterizing turbulent jet properties of radial discharge impeler: the zone of flow establishment (ZFE) and ZEF," AIChE Journal, 2013.

2. R. Ben-Nun and M. Sheintuch, "Charcterizing turbulent jet properties of radial discharge impeller: potential core, spreading rate and averaged flow field parameters," in Proceedings of the 9th European Congress of Chemical Engineering, The Hague, The Netherlands, April 2013.

3. S. Yeoh, G. Papadakis, and M. Yianneskis, "Numerical simulation of turbulent flow characteristics in a stirred vessel using the LES and RANS approaches with the sliding/deforming mesh methodology,"Chemical Engineering Research and Design, vol. 82, no. 7, pp. 834–848, 2004.

4. J. B. Joshi, N. K. Nere, C. V. Rane et al., "CFD simulation of stirred tanks: comparison of turbulence models. Part I: radial flow impellers," Canadian Journal of Chemical Engineering, vol. 89, no. 1, pp. 23–82, 2011.

5. J. B. Joshi, N. K. Nere, C. V. Rane et al., "CFD simulation of stirred tanks: comparison of turbulence models (Part II: axial flow impellers, multiple impellers and multiphase dispersions)," Canadian Journal of Chemical Engineering, vol. 89, no. 4, pp. 754–816, 2011.

6. M. Coroneo, G. Montante, A. Paglianti, and F. Magelli, "CFD prediction of fluid flow and mixing in stirred tanks: numerical issues about the RANS simulations," Computers and Chemical Engineering, vol. 35, no. 10, pp. 1959–1968, 2011.

7. A. Barker, R. D. Laroche, M. H. Wang, and R. V. Calabrese, "Sliding mesh simulation of laminar flow in stirred reactors," Chemical Engineering Research and Design, vol. 75, no. 1, pp. 42–44, 1997.

8. J. Gimbun, C. D. Rielly, Z. K. Nagy, and J. J. Derksen, "Detached eddy simulation on the turbulent flow in a stirred tank," AIChE Journal, vol. 58, no. 10, pp. 3224–3241, 2012.

9.   J. Derksen and H. E. A. Van Den Akker, "Large eddy simulations on the flow driven by a Rushton turbine," AIChE Journal, vol. 45, no. 2, pp. 209–221, 1999.

10.   J. Derksen, "Long-time solids suspension simulations by means of a large-eddy approach," Chemical Engineering Research and Design, vol. 84, no. 1, pp. 38–46, 2006.

11.   A. Bakker and L. M. Oshinowo, "Modelling of turbulence in stirred vessels using large eddy simulation,"Chemical Engineering Research and Design, vol. 82, no. 9, pp. 1169–1178, 2004.

12.   M. Jahoda, M. Mošt k, A. Kukuková, and V. Macho , "CFD modelling of liquid homogenization in stirred tanks with one and two impellers using large eddy simulation," Chemical Engineering Research and Design, vol. 85, no. 5, pp. 616–625, 2007.

13.   Z. Li, M. Hu, Y. Bao, and Z. Gao, "Particle image velocimetry experiments and large eddy simulations of merging flow characteristics in dual Rushton turbine stirred tanks," Industrial and Engineering Chemistry Research, vol. 51, no. 5, pp. 2438–2450, 2012.

14.   Z. Li, Y. Bao, and Z. Gao, "PIV experiments and large eddy simulations of single-loop flow fields in Rushton turbine stirred tanks," Chemical Engineering Science, vol. 66, no. 6, pp. 1219–1231, 2011

15.   J. J. Gillissen and H. E. Van den Akker, "Direct numerical simulation of the turbulent flow in a baffled tank driven by a Rushton turbine," AIChE Journal, vol. 58, no. 12, pp. 3878–3890, 2012.

16.   V. Kolá , P. Filip, and A. Curev, "Hydrodynamics of radially discharging impeller stream in agitated vessels," Chemical Engineering Communications, vol. 27, no. 5-6, pp. 313–326, 1984.

17.   J. Drbohlav, I. Fo t, K. Máca, and J. Ptá ek, "Turbulent characteristics of discharge flow from turbine impeller,"

Collection of Czechoslovak Chemical Communications, vol. 43, no. 12, pp. 3148–3161, 1978.

18. I. Fo t, H. O. Möckel, J. Drbohlav, and M. Hrach, "The flow of a liquid in a stream from the standard turbine impeller," Collection of Czechoslovak Chemical Communications, vol. 44, no. 3, pp. 700–710, 1979.

19. A. Obeid, I. Fo t, and J. Bertrand, "Hydrodynamic characteristics of flow in systems with turbine impeller," Collection of Czechoslovak Chemical Communications, vol. 48, no. 2, pp. 568–577, 1983.

20. J. Talaga and I. Fo t, "The velocity field in the discharge stream from a rushton turbine impeller," inProceedings of the 14th European Conference on Mixing, Warszava, Poland, September 2012.

21. B. C. Venneker, J. J. Derksen, and H. E. A. Van den Akker, "Turbulent flow of shear-thinning liquids in stirred tanks-The effects of Reynolds number and flow index," Chemical Engineering Research and Design, vol. 88, no. 7, pp. 827–843, 2010.

22. W.Bujalski, A. W. Nienow, S. Chatwin, and M. Cooke, "The dependency on scale of power numbers of Rushton disc turbines," Chemical Engineering Science, vol. 42, no. 2, pp. 317–326, 1987.

23. K. R. Beshay, J. Krat na, I. Fo t, and O. Br ha, "Power input of high-speed rotary impellers," Acta Polytechnica, vol. 41, no. 6, pp. 18–23, 2001.

Chapter 2

Engineering Parameters in Bioreactor's Design: A Critical Aspect in Tissue Engineering

Nasim Salehi-Nik[1,2], Ghassem Amoabediny[1,2], Behdad Pouran[1,2], Hadi Tabesh,[3] Mohammad Ali Shokrgozar[4], Nooshin Haghighipour[4], Nahid Khatibi,[1,2] Fatemeh Anisi[1,2], Khosrow Mottaghy[3], and Behrouz Zandieh-Doulabi[5]

[1]Department of Chemical Engineering, Faculty of Engineering, University of Tehran, Tehran, Iran

[2]Department of Biomedical Engineering, Research Center for New Technologies in Life Science Engineering, University of Tehran, Tehran, Iran

[3]Institute of Physiology, Medical Faculty, RWTH Aachen University, 52074 Aachen, Germany

[4]National Cell Bank, Pasteur Institute of Iran, Tehran, Iran
[5]Gustav Mahlerlaan 3004, 1081 LA Amsterdam, The Netherlands

ABSTRACT

Bioreactors are important inevitable part of any tissue engineering (TE) strategy as they aid the construction of three-dimensional functional tissues. Since the ultimate aim of a bioreactor is to create a biological product, the engineering parameters, for example, internal and external mass transfer, fluid velocity, shear stress, electrical current distribution, and so forth, are worth to be thoroughly investigated. The effects of such engineering parameters on biological cultures have been addressed in only a few preceding studies. Furthermore, it would be highly inefficient to determine the optimal engineering parameters by trial and error method. A solution is provided by emerging modeling and computational tools and by analyzing oxygen, carbon dioxide, and nutrient and metabolism waste material transports, which can simulate and predict the experimental results. Discovering the optimal engineering parameters is crucial not only to reduce the cost and time of experiments, but also to enhance efficacy and functionality of the tissue construct. This review intends to provide an inclusive package of the engineering parameters together with their calculation procedure in addition to the modeling techniques in TE bioreactors.

INTRODUCTION

Tissue engineering aims to generate three-dimensional (3D) artificial tissues. Its consequential task is to regenerate human tissue or develop cell-based substitutes for tissue in order to restore, reconstruct, or improve tissue functions [1, 2]. Achieving biological and mechanical functionality of the newly formed tissue is paramount for tissue engineered structures. Yet current research often focusses on form rather than function. Regeneration of functional organs demands intensive researches and studies in every aspect of TE [3], since creating a functional tissue requires the efficient growth of various types of cells on a single 3D structure

[2]. Bioreactors can aid the production of functional 3D tissues as follows: (1) by maintaining a desired uniform cell concentration within the scaffold during cell seeding, (2) by controlling microenvironmental parameters (e.g., temperature, pH, pressure, oxygen tension, metabolites, regulatory molecules, shear stress, and electrical pacing) and aseptic parameters (e.g., feeding, waste removal, and sampling), (3) by facilitating mass transfer [4–7], and more importantly (4) by allowing for automated processing steps.

Moreover, each special type of tissue structure and production procedure (e.g., skin, bone, blood vessel, cartilage, and myocardium) necessitates a unique kind of bioreactor design which requires both biological and engineering conditions to be addressed along with reliability, reproducibility, scalability, and safety issues [8–10].

In this review, key technical challenges between biological parameters and engineering parameters are recognized along with an overview of present mathematical modeling and monitoring of tissue growth carried out in the Research Center for New Technologies in Life Science Engineering at University of Tehran (UTLSE) to help deal with ongoing challenges.

ENGINEERING PARAMETERS IN TE BIOREACTOR DESIGN

Generally, the major responsibilities of a bioreactor are to provide a biomechanical and a biochemical environment that controls nutrient and oxygen transfer to the cells and metabolic products from the cells [11–13]. Mass transfer problems (e.g., oxygen and nutrient supply and removal of toxic metabolites) must always be taken into account. The size of most engineered tissues is limited as they do not have their own blood system and the cells are only nourished by diffusion [14, 15]. Since tissue constructions should have larger dimensions to become functional, mass transfer limitation can be considered as one of the greatest engineering challenges [1]. Moreover, biomechanical stimuli such as shear stress can be applied throughout the bioreactor by means of culture medium flow [10, 16]. In this condition, nutrient and waste transfer are automatically regulated by the flow of the growth medium. Other types of mechanical stimuli can also be applied to tissue constructs using a bioreactor, including axial compression or tensile forces [11].

Although biomechanical stimuli have many advantages for tissue engineering, mechanical stimuli can also induce tissue degradation, by alterations in the synthesis of matrix [16]. All in all, the response of some types of cells to mechanical stress causes radical changes to the tissue structure and composition which leads to alterations in tissue functionality.

In the following section, some of the engineering parameters which help providing physical stimulation to TE constructs in order to enhance tissue formation and their concomitant challenges are specified.

Mass Transfer through Bioreactors

The major obstacle that hinders practical application of 3D cell seeded constructs is mass transfer [5]. After distributing cells throughout porous scaffolds, a key challenge is the maintenance of cell viability, especially within the interior of the construct during prolonged culture. Nutrients, oxygen, and regulatory molecules have to be efficiently transferred from the bulk culture medium to the tissue surfaces (i.e., external mass transfer) as well as to the interior cells of the tissue construct (i.e., internal mass transfer). In addition, metabolites and CO_2 are to be removed from the cells within the tissue to the bulk medium. While external mass transfer rates depend primarily on hydrodynamic conditions in a bioreactor, internal mass transfer rates may depend on a combination of diffusion and convection mechanisms (typically induced by medium perfusion or scaffold deformation). Internal mass transfer depends strongly on the scaffold's structure and porosity, the overall cell or scaffold construct size, and the diffusion rate through the biomaterial [17, 18].

Improving the scaffold design will aid efficient mass transfer. For example, a laminar flow within tubular structures located within a scaffold may be beneficial for the generation of large TE constructs but requires the development of advanced bioreactor systems.

Amongst mass transfer mechanisms stated previously, oxygen transfer is a matter of the utmost importance due to poor solubility of oxygen in culture medium [9, 19, 20]. In addition, the diffusive penetration depth of oxygen within tissues in vivo is in the range of only 100 to 200 µm [19]. Thus, maintaining the balance between oxygen delivery to cells and their oxygen consumption is critical, considering this diffusive distance. Therefore, the oxygen tension adjustment is a critical matter in the design process of any bioreactor [21].

In applications germane to TE, the oxygen demand will fluctuate each time. During the initial expansion phase, cell density increases with time, and consequently, the overall demand for oxygen also increases. Cells may change from a proliferative state to the state of differentiation during the later stages of the culture. This change has implications for oxygen transfer, since proliferating cells typically have a higher oxygen demand per cell than differentiating cells [2]. Therefore, during the differentiation phase, the oxygen demand is likely to decline gradually.

A culture can be aerated by one, or a combination, of the following methods: surface aeration, direct sparging, indirect and/or membrane aeration (diffusion), medium perfusion, increasing the partial pressure of oxygen, and increasing the atmospheric pressure [22]. The transport of dissolved oxygen in a bioreactor occurs in three regions as follows:

- Bulk fluid phase of the bioreactor (global mass transfer).
- From the bulk to the surface of the aggregated cells (internal mass transfer).
- Through the aggregated cells (external mass transfer).

In the first step, at the gas-liquid interface, the rate of oxygen entering the medium is limited by the relatively low solubility of oxygen in aqueous medium. The scalar concentration distributions in the vessel for the global mass transfer depend on the flow field of the vessel and the net rate of consumption or production [23, 24]. Therefore, the oxygen concentration in the fluid experienced by the cells is a result of the balance between the oxygen delivery across the medium layer called the oxygen transfer rate (OTR) and the rate of oxygen consumption by cells named the oxygen uptake rate (OUR). Therefore, the oxygen concentration can be ten times lower as one would anticipate based on the equilibrium within the gas phase [25]. Oxygen availability has vigorous effect on cell culturing kinetics. For instance, increasing the amount of dissolved oxygen (DO) which can be done by increasing the OTR may lead to improve secondary metabolism too. The rate of OTR highly affects the liquid phase mass transfer coefficient ($k_L a$) and, then, the productivity. Therefore, it is essential to determine the DO level in the bioreactor [24, 26–28].

There are different methods for assessing the amount of oxygen delivered from the air to the culture environment (Table 2). The sulfite system is one in which transformed oxygen content from air to the aqueous solution is determined by means of the oxidation of sodium sulfite to sodium sulfate by oxygen. It could characterize the completing point of the reaction by means of a pH indicator, since the sulfate ions have more acidic activity than the sulfite ones. This method was applied in the presence of cobalt catalyst for determining the OTR and for studying the function of a perfusion bioreactor designed by UTLSE. It was concluded that oxygen delivery is appropriate and the bioreactor readily supplies the minimum required oxygen of the various cells. By considering the calculated OTR_{max} of 0.012 mol/L/hr and the largest $k_L a = 0.02/s$, calculations showed that bioreactor supplies the required oxygen of culturing more than 10^{10} CHO cells in the 80 mL culturing volume [29].

Mechanical Stimulation

The field of TE gradually recognizes the importance of mechanical stimuli (e.g., mechanical compression, mechanical stretch, hydrodynamic pressure, and fluid flow) in the maturation of organs [5]. Mechanical stimulation is one of particular interest for musculoskeletal tissue engineering, cartilage formation, and cardiovascular tissues [30–35]. Mechanical interactions during tissue growth, between different components, that is, cells, water, and scaffold material, can determine whether cells form cell aggregates or disperse throughout the scaffold [36–38]. Selection of optimal physical parameters is complicated by a variety of cell types, scaffolds, forces, applied regimes, and culture medium available.

Cells in aggregates are exposed to higher shear stresses than single cells due to their large particle diameter [39]. It is widely accepted that shear stress has a dominant impact on tissue function and viability. Different values are reported for the maximal sustainable shear stress for different types of cells [40, 41]. Indeed, high shear stress on the surface of the scaffold, caused by a flow of fluid, can peel off attached cells and in this condition, tissue growth is significantly slower compared with static cultures.

Simply, orientation and function of the cells is affected by fluid flow shear stress. Shear stress is a particular interesting stimulus for mammalian cell cultures because many cell types are responsive to shear stress [42–45]. For instance, it was observed that shear stress affected endothelial cell proliferation and oriented them toward flow direction [31]. There are many qualitative means for investigating fluid flow, which are summarized in Table 3.

In addition, the secretion of biological factors by stem cells can be increased by biomechanical forces. Therefore, it is important to acquire an understanding of the mechanisms by which hemodynamic forces are detected and converted into a sequence of biological responses within the cells [46]. For instance, changes in pressure or shear stress induce the rapid release of nitric oxide (NO) from the vascular endothelium [47–49]. Studies at UTLSE in a simple parallel plate flow chamber showed that NO production by Human umbilical vascular endothelial cells (HUVECs) is fluid shear stress rate dependent (data not shown).

In fact, the determination of how mechanical forces can be utilized is a challenge for bioreactor design in order to reach the proper environment necessary to produce the desired tissue engineered product. Pulsatile perfusion bioreactors integrated with elastic polymeric scaffolds enhance development and differentiation of small tissue engineered blood vessels [50–53]. Furthermore, custom-designed bioreactors utilizing biaxial strain for the mechanical stimulation of skeletal tissues were developed [54, 55].

Electrical Stimulation

In addition to mechanical stimuli commonly arising in tissue engineering context, electrical stimulation or even combined approaches incorporating electrical/mechanical cues need to be provided in vitro for obtaining an appropriate functionality of engineered tissue. Electrical stimuli are currently mainly applied in the field of cardiac tissue engineering to regenerate the infarcted area after heart failure [56, 57]. Radisic et al. [58] showed that electrical waves in a square form with frequency of 1 Hz and power of 5 V/cm can induce contractile properties in cardiac TE constructs. The disruption of regularity of ions in an electrically affected construct leads to redistribution of charge which can then alter the pH gradient in the media which can be used to tailor specifically enhanced cellular function [59]. Finally, electrical pacing associated with mechanical cues in the culture when applied to the electrospun cardiac constructs resulted in better alignment, elongation, and upregulation of cardiac proteins compared with static cultures [60].

COMPARISON BETWEEN DIFFERENT TYPES OF TE BIOREACTORS BASED ON ENGINEERING PARAMETERS

Bioreactors that are currently widely used in TE are static and mixed flasks, rotating wall, and perfusion bioreactors. These bioreactors offer three distinct flow conditions (static, turbulent, and laminar), and hence a different rate of nutrient supply to the surface of tissue construct [24]. They also differ in mass transfer and shear stress rates experienced by the cultured cells. Table 1 compares engineering parameters of different TE bioreactors.

Table 1: Comparison of engineering parameters in different TE bioreactors

Bioreactor type	General descriptions	Mass transfer mechanism	Shear stress	Special usage	Tissue	Considerations
Static culture	Batch culture with no flow of nutrient	Diffusion (high)	Very low	Cell proliferation	—	Homogeneous structure of cell constructs and nutrient diffusion limitations
Stirred flasks	Magnetically stirring of medium	Convection (high)	High	Dynamic seeding of scaffolds	Cartilage	Appropriate scaffold and balance between increasing mass transfer and modulating shear stresses

Rotating wall	Rotating at a speed so the constructs in the reactor are maintained "stationary" in a state of continuous free fall	Convection (high)	Low	Tissue constructs which need dynamic laminar flow	Cartilage, bone and skin	Operating conditions (e.g., speed of rotating) especially for growing large tissue mass
Perfusion	Flow of medium over or through a cell population or bed of cells	Convection (moderate) and diffusion (high)	Moderate	Tissues physico-chemicaly and environmentally relevant to human tissues	Epithelial cells, intestinal, bone, cartilage, and arteries	Seeding and attachment of human cells especially within the scaffold body

Table 2: Methods of measuring oxygen transfer rate

Measurement method	Basis of the method	Pros	Cons	Ref.
Sulfite oxidation method	Monitoring pH changes during the oxidation of sodium sulfite to sodium sulfate controlled by oxygen depletion rate	(i) Simple and low cost (ii) Can be used for the determination of the interfacial area between gas and liquid (iii) Being accurate for minivolumes of less than 1 mL	(i) The kinetics of the homogeneous catalytic chemical reaction should be known (ii) Limited accuracy by visually determination of color change (iii) High salt concentration (usually $0.5 \, mol \cdot L^{-1}$) reduces the maximum solubility of oxygen (iv) Not appropriate in large scale bioreactors (v) High surface tension causes the underestimation of potentially achievable OTR	[72–75]
Dynamic method	Monitoring the dissolved O_2 concentration during the aeration of the system	(i) Consistent measurement (ii) Does not depend on a zero or reference measurement	Requiring a rapidly responsive, sterilizable, dissolved oxygen probe (ii) Limited application for minititerplates (MTPs) (iii) Not costly favorable	[76, 77]
Optical method	Monitoring the color changes during the sulfite oxidation reaction using a pH sensitive dye (e.g., bromothymol blue)	No need for pH electrode which frequently disturbs the hydrodynamics	Not accurate due to being time dependent of the color shift which indicated the time of the oxidation reaction	[78–80]

Gassing-out method	Monitoring k_La by direct measurement of the rate of increasing dissolved oxygen concentration, after neutralizing the system by flushing nitrogen through the vessel to achieve an oxygen-free solution	(i) Can be applied to different media (for investigating the effect of media composition on oxygen mass transfer) (ii) Does not involve chemical reactions that could impact the measurement precision and the liquid film resistance	A nonrespiring system which is not in exact correspondence to real culturing conditions	[28, 81,82]
RAMOS (intermittent online) method	Monitoring OTR by periodically repeating an automated measuring cycle composed of a measuring phase and a rinsing phase	(i) Online monitoring system (ii) Can be used simultaneously for measuring OTR in 6–12 parallel vessels	(i) Large amount of sample required for measuring (ii) Not applicable for small volumes (microliters)	[83–85]
Exhaust gas analyzer (continuous online) method	Calculating the OTR by specifying the oxygen concentration difference between the inlet gas stream (O_2, in) and the outlet gas stream (O_2, out) using magnetomechanical exhaust gas analyzer (EGA)	(i) Continuous method (ii) Can be used to measure OTR in one to five parallel culture vessels	Only applicable in high volume bioreactors	[86, 87]
Respirometer (offline) method	Measuring of decreasing dissolved oxygen concentration with time after aerating the culture vessel	Can be used for bioreactors of any shape	Difficult manual handling	[88, 89]

Table 3: Methods of measuring flow

Measurement method	Basis of the method	Pros	Cons	Ref.
Particle image velocimetry (PIV), including micro-PIV (µPIV)	Monitoring the displacement of small seeded particles in a region of interest of fluid medium via double-pulsed laser beam	(i) Can be used through an in vitroinvestigation (ii) Noninvasive method (iii) High spatial resolution (iv) Simultaneously determination of velocities of two different phases without disturbing the flow	(i) Almost impossible for in vivo experiments (ii) Requiring undistorted optical access to the area of interest for both an excitation laser and an imaging system (iii) Limited of temporal resolution (iv) Requiring appropriate particles to eliminate the differences between solid particles and local fluid velocities	[90–94]
Holographic PIV (HPIV)	Record the particle image field using a reference beam to project the hologram, followed by a 2D plane detector moved through the projected hologram	(i) Can also record a 3D instantaneous flow field (ii) Being user friendly	(i) Reduction of speckle noise (ii) Not handling huge quantities of data (iii) Cannot extract 3D velocity in presence of large gradients/fluctuations (iv) Complexity of system (v) Requiring large depth of focus that affects the measurement accuracy	[93–96]

Particle tracking velocity (PTV)	Measuring particle velocities using video camera recording	(i) Easily determination of even small displacement of particles without confusing them with neighboring ones (ii) Measurement of velocity at the location of a particle, without requiring an averaging over a grid (compared to PIV)	(i) Requiring many individual particles to be reconstructed in space and identified in successive frames (ii) Lower spatial resolution (iii) Time consuming	[95–98]
Laser Doppler anemometry (LDA) or laser Doppler velocimetry (LDV)	Measuring of scattered laser light by particles that pass through a series of interference fringes (a pattern of light and dark surfaces)	(i) High spatial and temporal resolution (typically in the order of 1 kHz) (ii) Nonintrusive method (iii) No calibration required (iv) Recording one, two, or three velocity components simultaneously (v) Also applicable in reversing flows	(i) Cannot simultaneously measure the velocities of different phases (ii) Time consuming (iii) Difficult to analyzing the discrete data stream from the flow	[94–97,99]
Acoustic Doppler velocimeter (ADV)	Measuring the velocity of particles in a remote sampling volume based on the Doppler shift effect using one transmitter and three receivers	Simultaneously recording nine values with each sample: three velocity components, three signal strength values, and three correlation values	(i) Only suitable for flow conditions with relatively low turbulence level (ii) Required postprocessing of data	[99,100]

Holographic correlation velocimetry (HCV)	Measuring 3D velocity fields of a fluid at high speed combining a correlation-based approach with in-line holography	(i) Very efficient with regard to the use of light, as it does not rely on side scattering (ii) Very high quality system at a modest cost (iii) Appropriate for high-speed flows and low exposure times (iv) Simple calibration (v) Using relatively low powered lasers (vi) Direct measurement of the velocity field at all depth locations (vii) Nonintrusive technique especially in cell culture procedures	Requiring a separate method to extract velocity data from holographic images	[94, 95]

Although static culture is simply designed and operated, there are nutrient diffusion limitations with large constructs since both external and internal mass transfer are undertaken by diffusion [9, 11, 22]. Statically cultured constructs often have a heterogeneous structure and composition, including a necrotic central region and dense layers of viable cells encapsulating the construct outer edge [17]. This condition appears due to concentration gradients, with local depletion of nutrients and accumulation of waste materials [18].

Cell survival and assembly on many surfaces of engineered tissues can be improved by construct cultivation instirred flask bioreactors [61–65]. Within such flasks, scaffolds are attached to needles hanging from the lid of the flask for dynamic seeding. Convective flow, generated by a magnetic stirrer bar, allows continuous mixing of the medium surrounding the construct [24]. This environment improves nutrient diffusion and promotes cell proliferation throughout the constructs in comparison to static condition. However, the shear forces acting on the constructs are heterogeneous, which prevents homogenous tissue development [11].

In order to enhance external mass transfer under a laminar flow condition, the tissue engineered constructs can be cultivated in rotating wall bioreactors [63, 65–67]. Dynamic laminar flow of rotating bioreactors generally improves properties of the peripheral tissue layer. Also, in such bioreactors, no fibrous capsule is formed, but the limitations of the diffusional transfer of oxygen to the construct interior still remain [24]. As compared to the turbulent flow within stirred flasks, the dynamic laminar flow in rotating wall vessels contributes to reduced levels of shear stress experienced by cells on the construct. Amongst other, this aides the formation of cartilaginous tissues containing higher amounts of more uniformly distributed glycosaminoglycans (GAG) and collagen [18, 68].

In addition, a key point to note is that convective transfer around and through an engineered tissue at the proper flow rate can dissipate gradients of nutrients and maintain tissue mass [69]. In a novel strategy, Yu et al. [70] mixed microspheres of different densities in order to vary and modify flow velocity within a scaffold through the rotating wall bioreactor. Compared to static three-dimensional controls, culturing rat primary calvarial cells under dynamic flow conditions in a rotating system reveals a more uniform distribution of cells in the scaffold interior and also enhances phenotypic protein expression and recuperates mineralized matrix synthesis. In addition, Zhang et al. [71] recognized that scaffolds seeded by human fetal mesenchymal stem cell (hfMSC) reached cellular confluence earlier with greater cellularity and also conserved high cellular viability in the core of them compared to a static culture.

Perfusion bioreactors are used in order to force culture medium through the pores of solid porous 3D scaffolds, thereby enhancing nutrient transport and providing mechanical stimuli to the cells (e.g., [63, 65, 101–104]). In such systems, oxygen and nutrients are supplied to the construct interior by both diffusion and convection. The flow rate can be optimized with respect to the limiting nutrient, which is mostly oxygen due to its low solubility in culture medium [24, 105]. Perfusion of chondrocyte-seeded scaffolds was reported to elevate GAG synthesis and retention within the extracellular matrix (ECM) [106], as well as a uniform distribution of viable human chondrocytes. A perfusion system can provide a well-defined physicochemical culture environment which has great potential to generate cartilage grafts [68] or vascular grafts of clinically relevant size [107, 108]. Bioreactors that perfuse the culture medium directly through the pores of a scaffold enhance mass transfer rate not only at the construct periphery but also within the internal pores. This can potentially eliminate mass transfer limitations. Perfusion bioreactors can offer greater control of mass transfer than other conventional systems but the potential for flow to follow a preferential path through the construct still remains a problem. This phenomenon happens particularly for scaffolds with a wide pore size distribution and nonuniformly developing tissues, leaving some regions poorly nourished, while others are perfused strongly.

It is confirmed that cartilage-like matrix synthesis by chondrocytes, chondrocyte growth, and differentiation and deposition of mineralized matrix by bone cells are enhanced by direct perfusion bioreactors [109]. It is worth to notice that the flow rate in the microenvironment of cells is to a great extent responsible for the changes of medium perfusion. Therefore, to optimize a perfusion bioreactor for tissue engineering applications, the balance between the extent of nutrient supply, the transport of metabolites to and away from cells, and the fluid-induced shear stress effects on cells located at the surface and in the porous structures of the scaffold should be considered [17, 21, 105].

In order to gain a better understanding on how physical factors modulate tissue development, it is necessary to integrate bioreactor studies with quantitative analyses and computational modeling of changes in mass transfer and physical forces experienced by cells [17].

MATHEMATICAL MODELING OF ENGINEERING PARAMETERS

Mathematical modeling in terms of fundamental physical and biochemical mechanisms can be used to justify experimental results and determine future research directions [110–113]. Relatively few mathematical modeling studies have focused on bioreactor culture of cell-seeded porous structures for TE [114, 115].

In the first stage, numerical simulation plays an important role in prediction of the global dynamic response in different parts of bioreactors. Moreover, numerical evaluation provides insight into local hydrodynamic changes in tissue constructs in order to generate quantitative anticipation of the tissue development within a bioreactor system [116]. Finally, with the aid of recently available computational tools, variables (e.g., flow fields of a particular bioreactor design [117, 118], incorporation of the mechanics of the scaffold material [119], and the sufficiency of bioreactor cultures [117, 120, 121], shear stresses and mass transfer in scaffold-containing bioreactors [118, 122]) can be estimated.

As an example, Sengers et al. [3] in their review concentrated on the contribution of computational modeling as a framework to obtain an integrated understanding of key processes including nutrient transfer, matrix formation, dynamics of cell population, cell attachment and migration, and local mutual interactions between cells.

Nutrient and Mass Transfer

The amount of delivered oxygen is a significant factor in designing the cell culture bioreactors. One major obstacle preventing proper understanding of oxygen tension in TE constructs is a lack of mathematical models that can predict which parameters are beneficial for avoiding oxygen limitation and increasing oxygen diffusion across serial resistances [114, 118, and 121]. This problem was resolved at UTLSE by applying traditional convective mass transfer models combined with Maxwell-Stefan diffusion mass transfer equation.

The reliability of the model can be examined by comparing the model results obtained with sulfite experiments done with four geometries of shake flasks (Figure 1).

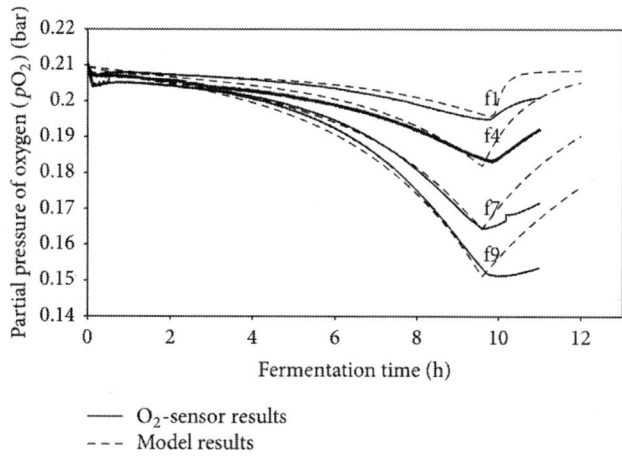

Figure 1: Comparison between unsteady state model and experimental results for the partial pressure of oxygen in the headspace of the ventilation flasks f1, f4, f7, f9 (sterile plug dimensions in f1 < f4 < f7 < f9) is obtained for the fermentation of C. glutamicum DM 1730 on 10 g/L glucose and 21 g/L MOPS (VL = 10 mL, n = 400 rpm, T = 30° C, do = 5 cm, Yx/s = 0.48, $Yx/o2$ = 53 g/mol, RQ = 1 where do, VL, Yx/s, $Yx/o2$, and RQ are shaking diameter, filling volume, yield of biomass with respect to substrate, yield of biomass with respect to oxygen, and respiration quotient, resp).

As can be seen from Figure 1, the value for the pO_2 is 0.2095 bar at the onset of the experiment. Oxygen partial pressure decreases over time as the chemical reaction proceeds. The flasks with the greater sterile plug dimensions represent lower mass transfer rates which resulted from hindered diffusion. This gives rise to a lower partial pressure of oxygen [84, 85, 123, and 124]. Yan et al. [114] developed a novel mathematical model to represent the glucose and oxygen distribution and the cell growth in a 3D cell-scaffold construct in a perfusion bioreactor. Numerical methods are employed to solve the equations involved, with a focus on investigating the effect of various factors such as culturing time, porosity, and flow rate, which are controllable in the scaffold fabrication and culturing process, on cell cultures.

Along these lines, Pisu et al. [125] proposed an improved description of oxygen consumption and GAG production by bovine chondrocytes, which is thoroughly related to cellular metabolism. The latter is simulated through appropriate population balance models which include cellular anabolic and catabolic rates.

Abdollah and Das [126] presented a general modeling framework to characterize nutrient (oxygen and glucose) transfer in a hollow fiber membrane bioreactor (HFMB) for bone tissue growth. The framework relied on solving coupled Navier-Stokes and the Maxwell-Stefan convection-diffusion-reaction equations. It is indicated that due to multicomponent interactions, mass severe transfer limitations may arise severely when inlet concentration of nutrients, molecular size of the solutes, and wall membrane thickness are increased.

Rivera-Solorio and Kleis [23] used a mathematical model to investigate the local mass transfer of dissolved oxygen to the surface of freely suspended cell aggregates in a bioreactor operating in microgravity. They simulated the mass transfer in systems in which cultured cells are attached to small microcarriers in a rotating bioreactor in simulated and real microgravity. Also, Yu et al. [127] evaluated oxygen transfer in a microbioreactor for animal cell suspension culture using the commercial software Fluent. They proposed two correlations in order to calculate the liquid-phase oxygen transfer coefficient and the minimum oxygen concentration in a microbioreactor, to provide insight into choosing the proper operating parameters in animal cell culture.

Fluid Flow

To better realize the effect of fluid flow during tissue regeneration, a number of studies using computational fluid dynamic (CFD) have been accomplished [128–134]. These CFD studies revealed detailed profile of pressure, velocity, flow fields, shear stresses, and oxygen transfer in tissue culturing chambers of various bioreactor designs. This is very useful for the design optimization of internal geometric configurations of bioreactors [116].

Lawrence et al. [16] explored the effect of reactor geometry on flow fields using the computational fluid dynamics software Comsol Multiphysics 3.4. The Brinkman equation was used to model the permeability characteristics within the chitosan porous structure. Results showed significant increase in pressure with reduction in pore size, which could limit the fluid flow and nutrient transport.

Subsequently, flow characteristics are analyzed using either Darcy's equation [135] or the Brinkman equation considered as an extension of Darcy's equation. The Brinkman equation accounts for both viscous and drag forces in the porous medium. It can be reduced to either Navier-Stokes equation or Darcy's law if forces become dominant. The Brinkman equation is as follows [16]:

$$\mu \nabla^2 u_s - \frac{\mu}{k} u_s = \nabla p, \qquad \nabla u_s = 0,$$

$$(1)$$

where k the permeability of the porous medium, u_s denotes the fluid superficial velocity vector, p the fluid pressure, and μ is the effective viscosity in the porous medium. Nonporous sections of a bioreactor were modeled as incompressible Navier-Stokes regions. The permeability of the porous medium (k) is a geometric characteristic of the porous structure at several length scales. The Navier-Stokes equation together with continuity equation provides an essential tool to investigate the mechanical behavior of fluid in shaken bioreactors.

Our research center began intensive studies of hydrodynamics applying CFD in shaken micro bioreactors including 24-well plates, shaken at various shaking frequencies. For instance, schematics of the liquid phase fraction and radial velocity profiles at 0.7 cm distance from the bottom plane are shown in Figures 2 and 3 at shaking frequency of 200 rpm.

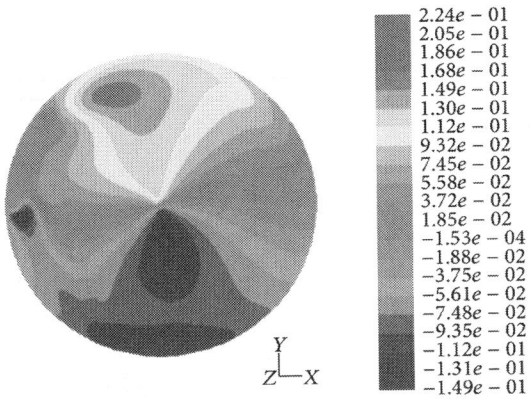

2.24e − 01	
2.05e − 01	
1.86e − 01	
1.68e − 01	
1.49e − 01	
1.30e − 01	
1.12e − 01	
9.32e − 02	
7.45e − 02	
5.58e − 02	
3.72e − 02	
1.85e − 02	
−1.53e − 04	
−1.88e − 02	
−3.75e − 02	
−5.61e − 02	
−7.48e − 02	
−9.35e − 02	
−1.12e − 01	
−1.31e − 01	
−1.49e − 01	

Contours of radial velocity (mixture) (m/s) (time = 3.0075e − 01)

Figure 2: Radial velocity distribution in a shaken 24-wells bioreactor that illustrates inhomogeneous map of radial velocity at the interface of liquid and air.

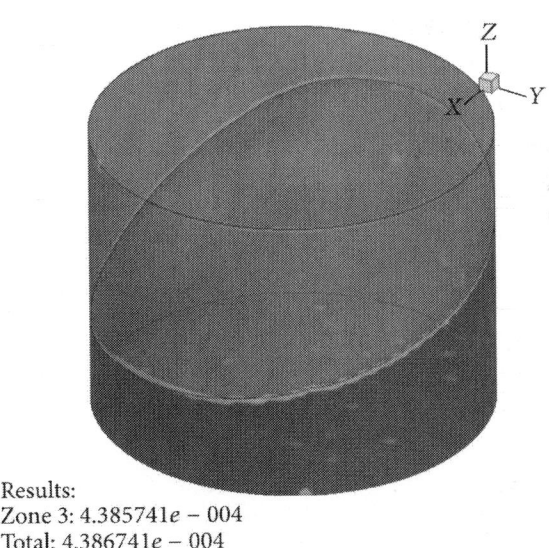

Results:
Zone 3: 4.385741e − 004
Total: 4.386741e − 004

Figure 3: Volume fraction distribution in a shaken 24-wells bioreactor that allows accurate prediction of gas-liquid interface area within the shaking bioreactor.

Output data from phase fraction simulation gave insight in the gas-liquid interfacial surface area, which then helps to determine the exact mass transfer coefficient ($k_L a$) values. Furthermore, the mean radial velocity at the interface provides a guideline for obtaining wall shear stress within the entire domain of the bioreactor.

The outcomes of shear stress simulation experiments confirm that the magnitude of this mechanical quantity rarely exceeds 1 Pa at the bottom of the plate. This value of shear stress can be withstood by most mammalian cells [136].

In addition, a novel flow chamber was developed in our research center to assess the effect of fluid flow on the efficiency of nutrient transport and the endothelial cell stability. This chamber exhibits the major features of a standard parallel flow bioreactor in which a circular silicon scaffold is centrally located. To accomplish this, CFD was used to discretize mathematical equations. Energy dissipation rate (EDR) and shear stress were plotted versus position in the cylinder at a flow rate of 75 mL/min (Figure 4).

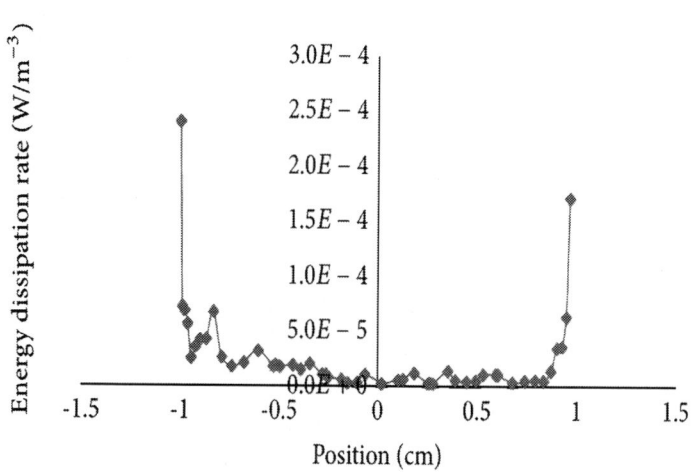

(a)

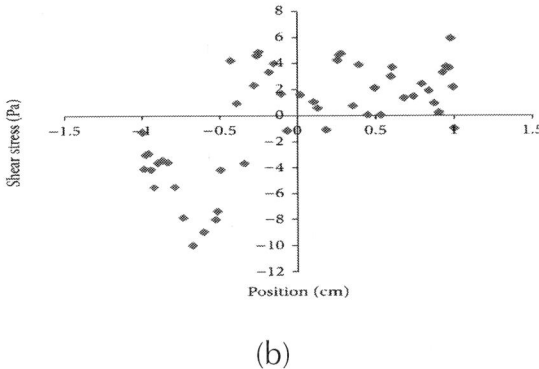

(b)

Figure 4: Energy dissipation rate (a) and shear stress distribution (b) versus radial position on a scaffold with radius of 1 cm that obviously represents safe generated shear stress on the scaffold for mammalian cell cultures.

For either plots of EDR and shear stress, a symmetrical pattern reveals that a homogeneous distribution of these mechanical characteristics of flow exists. However, for EDR data, some values deviate slightly between both sides of the cylinder because of a flow maldistribution which is due to a mild turbulence over the scaffold.

After completion of simulations, cell experiments were conducted for a 1 hr period. These experiments showed that at a volumetric flow rate of 75 mL/min, the cell viability and stability are maintained, but no specific cell orientation is present (Figure 5).

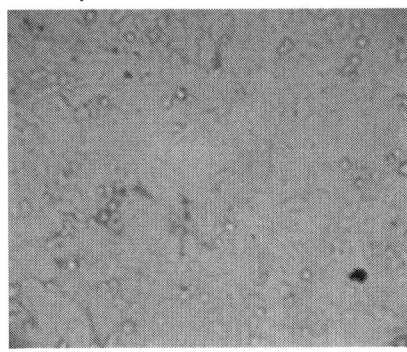

(a)

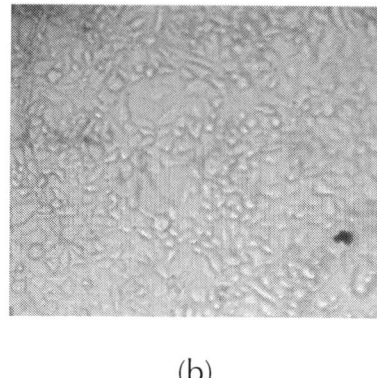

(b)

Figure 5: Schematics of cell morphology (a) before (b) after initiation of flow indicating flow assisted elongation of cells under continuous flow.

In general, computational fluid dynamics applications in bioreactor development can be extended to new designs such as a novel perfusion bioreactor developed at UTLSE.

In order to assess mechanical as well as oxygen characteristics of this novel perfusion bioreactor, scientists at UTLSE used CFD to determine fluid velocity as well as path lines features of the flow. Figure 6 further describes the computational attributes of the system.

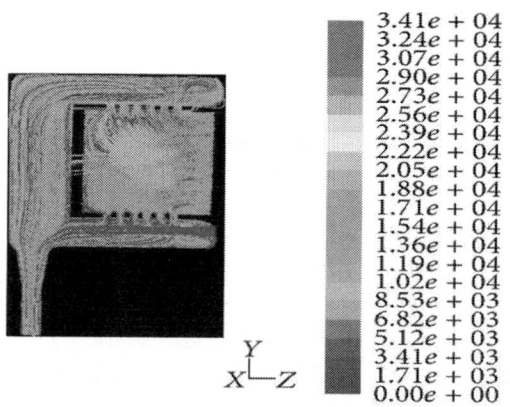

Path lines colored by particle ID

(a)

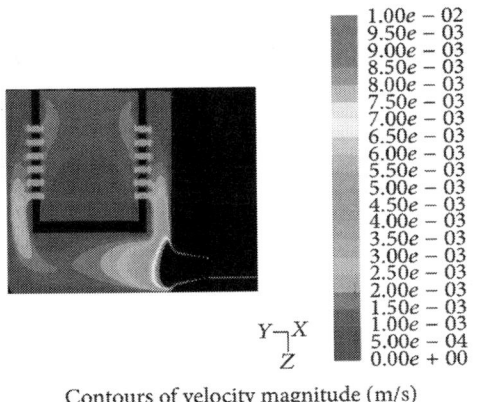

1.00e − 02
9.50e − 03
9.00e − 03
8.50e − 03
8.00e − 03
7.50e − 03
7.00e − 03
6.50e − 03
6.00e − 03
5.50e − 03
5.00e − 03
4.50e − 03
4.00e − 03
3.50e − 03
3.00e − 03
2.50e − 03
2.00e − 03
1.50e − 03
1.00e − 03
5.00e − 04
0.00e + 00

$Y\!\!-\!\!\urcorner X$
Z

Contours of velocity magnitude (m/s)

(b)

Figure 6: Demonstration of path lines to track the fluid particles within the bioreactor (a) and velocity magnitude to evaluate maximal shear stress in order to optimize shear stress distribution (b) in a perfusion bioreactor belonging to UTLSE.

The initial approximation of fluid flow dynamics attained with CFD is extremely beneficial in reducing time and costs of development of the bioreactor [29].

Yu et al. [137] applied a CFD model to simulate the flow and oxygen concentration fields in a micro bioreactor, in which a small magnetic bar was placed in a culture well to enhance the medium mixing. It was found that the hydrodynamic environment could be appropriate for animal cell culture when the microbioreactor operated at a stirrer rotating speed of 300 rpm and working volume of 4 mL.

Bilgen and Barabino [138] took advantage of CFD modeling to characterize the complicated hydrodynamic environment of a wavy-walled bioreactor applied for cultivation of tissue-engineered cartilage structures. They also analyzed the changes in the flow field when TE constructs are present in the bioreactor. The flow-induced shear stress experienced by engineered constructs cultivated in the wavy walled bioreactor was much lower than that of spinner flask. The radial or axial position of the constructs can modulate this shear stress. Lawrence et al. [16] used rectangular and circular bioreactors with three different inlet and outlet paradigms. By the use of CFD, geometries were simulated in two cases, with and without the presence of a porous structure. Residence time distribution analysis using the change of a tracer within a bioreactor revealed nonideal fluid distribution characteristics. The result represented a significant increase in pressure with a decrease in pore size, which could lead to low fluid flow and nutrient transfer limitation.

Cell Growth, Proliferation, and Viability

Chung et al. [111] developed a mathematical model for the static culture of cells grown on porous scaffolds. Results showed that the overall cell growth allows cells to spread more uniformly, while it prevents cells from competing for nutrients at the same site. They then described a mathematical model to examine the effects of medium perfusion on the cell-scaffold constructs [120]. They proposed a three-layer model, highlighting the enhancement of cell growth by medium perfusion. The model is quite detailed, involving a cell construct sandwiched between two fluid layers in order to mimic the culturing environment of direct perfusion. Although the model is valuable in developing engineered cell constructs, the enormous number of essential formulas and boundary conditions make the model cumbersome. Therefore, a compact mathematical model was to describe cell growth within a porous scaffold under direct perfusion. Neglecting the two fluid regions sandwiching the scaffold, the model contains only the scaffold region for computational purposes [110].

Shakeel [118] in his thesis developed a model which describes the key features of the tissue engineering processes such as the interaction between the cell growth, variation of material porosity, flow of fluid through the material, and delivery of nutrients to the cells. The fluid flow through the porous scaffold and the delivery of nutrients to the cells was modeled by Darcy's law and the advection-diffusion equations, respectively. For modeling the cell growth, a nonlinear reaction diffusion system was used. The results show that the distribution of cells and total cell number in the scaffold depends on the initial cell density and porosity of the scaffold.

A unique set of dynamical mathematical models was used to accurately predict metabolite and cell concentration in an aerated miniaturized shaking bioreactor at UTLSE. The major advantage of such a mathematical model is that it provides a robust tool to solve complicated oxygen transfer which unfavorably hampers metabolite production in bioprocesses.

The combination of equations which make a link between liquid phase oxygen concentration and rate of oxygen uptake with governing equations of cell concentration should be primarily solved to attain oxygen transfer rate (OTR) with respect to the course of time. Figure 6 suggests that as the model microorganism is undergoing accelerating growth, the oxygen transfer rate increases until the growth is inhibited and consequently the OTR falls down significantly. Figure 7 illustrates the comparison between the model and experimental results for a model microorganism, which suggest that a minor discrepancy between their model and the results exist [139].

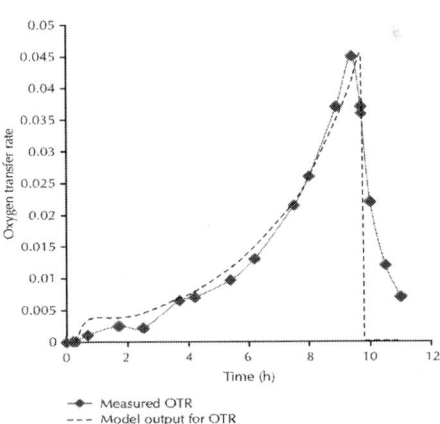

Figure 7: Comparison of OTR resulting from model and from experiments for a specific aerobic microorganism. The plot provides evidence of the proximity of OTR values between experimental and simulation results and of the efficacy of the simulation efforts.

CONCLUSIONS

Engineering parameters occurring in a bioreactor are of equal importance as biological parameters and should therefore be investigated thoroughly in order to optimize outcomes of TE strategies. Internal and external mass transfer (e.g., oxygen, nutrient, and waste materials transfer) as well as mechanical stimulation (e.g., fluid flow and shear stress) should be monitored online. Between different types of bioreactors, the "Perfusion Bioreactor" is the most convenient for animal cell cultures on a solid porous scaffold. Perfusion bioreactors offer both convection and diffusion and can provide nearly in vivo physiochemical and environmentally stimuli for engineered tissue constructs.

The operating conditions for diverse bioreactors can be very different per experiment. Therefore, it is essential to use mathematical equations and modeling techniques to simulate the optimal operating conditions in order to predict the best outcomes. Using the Brinkman equation along with powerful CFD codes can provide for investigating the effects of engineering parameters on the outcome of biological experiments. In this way, the efficacy of bioreactors, which is very low at present, can be optimized.

ACKNOWLEDGMENTS

The great efforts of all the students in the group of Tissue Engineering at the Department of Biomedical Engineering, UTLSE, who did an extensive research in this field, especially N. Noormohammadi, M. H. Gholami, B. Zamiri, M. Badv, Sh. Falamarzian, and P. Banikarimi are gratefully acknowledged.

REFERENCES

1.   R. Portner, S. Nagel-Heyer, C. Goepfert, P. Adamietz, and N. M. ¨ Meenen, "Bioreactor design for tissue engineering,"

Journal of Bioscience and Bioengineering, vol. 100, no. 3, pp. 235–245, 2005.

2. M. Ellis, M. Jarman-Smith, and J. B. Chaudhuri, "Bioreactor systems for tissue engineering: a four-dimensional challenge," in Bioreactors For Tissue Engineering: Principles, Design and Operation, M. Al-Rubeai and J. B. Chaudhuri, Eds., pp. 1–18, Springer, 2005.

3. B. G. Sengers, M. Taylor, C. P. Please, and R. O. C. Oreffo, "Computational modelling of cell spreading and tissue regeneration in porous scaffolds," Biomaterials, vol. 28, no. 10, pp. 1926–1940, 2007.

4. M. Radisic, H. Park, and G. Vunjak-Novakovic, "Cardiac-tissue engineering," in Principles of Tissue Engineering, R. Lanza, R. Langer, and J. P. Vacanti, Eds., Academic Press, 3rd edition, 2008.

5. J. J. Pancrazio, F. Wang, and C. A. Kelley, "Enabling tools for tissue engineering," Biosensors and Bioelectronics, vol. 22, no. 12, pp. 2803–2811, 2007.

6. R. I. Freshney, B. Obradovic, W. Grayson, C. Cannizzaro, and G. Vunjak-Novakovic, "Principles of tissue culture and bioreactor design," in Principles of Tissue Engineering, R. Lanza, R. Langer, and J. P. Vacanti, Eds., Academic Press, 3rd edition, 2008.

7. Y. I. Yang, D. L. Seol, H. I. Kim, M. H. Cho, and S. J. Lee, "Continuous perfusion culture for generation of functional tissue-engineered soft tissues," Current Applied Physics, vol. 7, no. 1, pp. e80–e84, 2007.

8. R. Portner and C. Giese, "An overview on bioreactor design, ¨ prototyping and process control for reproducible threedimensional tissue culture," in Culture of Cells For Tissue Engineering, G. Vunjack-Novakovic and R. Ian Freshney, Eds., pp. 53–78, John Wiley & Sons, 2006.

9. I. Martin, D. Wendt, and M. Heberer, "The role of bioreactors in tissue engineering," Trends in Biotechnology, vol. 22, no. 2, pp. 80–86, 2004.

10.   Y. Martin and P. Vermette, "Bioreactors for tissue mass culture: design, characterization, and recent advances," Biomaterials, vol. 26, no. 35, pp. 7481–7503, 2005.

11.   A. J. El Haj, M. A. Wood, P. Thomas, and Y. Yang, "Controlling cell biomechanics in orthopaedic tissue engineering and repair," Pathologie Biologie, vol. 53, no. 10, pp. 581–589, 2005.

12.   E. M. Bueno, B. Bilgen, R. L. Carrier, and G. A. Barabino, "Increased rate of chondrocyte aggregation in a wavy-walled bioreactor," Biotechnology and Bioengineering, vol. 88, no. 6, pp. 767–777, 2004.

13.   H.-C. Chen and Y.-C. Hu, "Bioreactors for tissue engineering," Biotechnology Letters, vol. 28, no. 18, pp. 1415–1423, 2006.

14.   H. Tabesh, G. Amoabediny, N. S. Nik et al., "The role of biodegradable engineered scaffolds seeded with Schwann cells for spinal cord regeneration," Neurochemistry International, vol. 54, no. 2, pp. 73–83, 2009. 12 BioMed Research International

15.   R. Y. Kannan, H. J. Salacinski, K. Sales, P. Butler, and A. M. Seifalian, "The roles of tissue engineering and vascularisation in the development of micro-vascular networks: a review," Biomaterials, vol. 26, no. 14, pp. 1857–1875, 2005.

16.   B. J. Lawrence, M. Devarapalli, and S. V. Madihally, "Flow dynamics in bioreactors containing tissue engineering scaffolds," Biotechnology and Bioengineering, vol. 102, no. 3, pp. 935– 947, 2009.

17.   D. Wendt, N. Timmins, J. Malda, F. Janssen, A. Ratcliffe, G. Vunjak-Novakovic et al., "Bioreactors for tissue engineering," in Tissue Engineering, C. van Blitterswijk, P. Thomsen, J. Hubbell, R. Cancedda, J. D. de Bruijn, A. Lindahl et al., Eds., pp. 484–506, 2008.

18.   P. Rolfe, "Sensing in tissue bioreactors," Measurement Science and Technology, vol. 17, no. 3, pp. 578–583, 2006.

19.   G. F. Muschler, C. Nakamoto, and L. G. Griffith, "Engineering principles of clinical cell-based tissue engineering," Journal

of Bone and Joint Surgery A, vol. 86, no. 7, pp. 1541–1558, 2004.

20. S. J. Wang and J. J. Zhong, "Bioreactor engineering," in Bioprocessing For Value-Added Products From Renewable Resources, S. T. Yang, Ed., Elsevier, 2007.

21. R. Depprich, J. Handschel, H.-P. Wiesmann, J. Jasche-Meyer, ¨ and U. Meyer, "Use of bioreactors in maxillofacial tissue engineering," British Journal of Oral and Maxillofacial Surgery, vol. 46, no. 5, pp. 349–354, 2008.

22. R. Eibl, D. Eibl, R. Portner, G. Catapano, and P. Czermak, Cell and Tissue Reaction Engineering, Springer, New York, NY, USA, 2008.

23. I. Rivera-Solorio and S. J. Kleis, "Model of the mass transport to the surface of animal cells cultured in a rotating bioreactor operated in micro gravity," Biotechnology and Bioengineering, vol. 94, no. 3, pp. 495–504, 2006.

24. J. Malda, M. Radisic, S. Levenberg et al., "Cell nutrition," in Tissue Engineering, C. van Blitterswijk, P. Thomsen, J. Hubbell et al., Eds., pp. 328–362, 2008.

25. F. Garcia-Ochoa and E. Gomez, "Bioreactor scale-up and oxygen transfer rate in microbial processes: an overview," Biotechnology Advances, vol. 27, no. 2, pp. 153–176, 2009.

26. R. Hermann, M. Lehmann, and J. Buchs, "Characterization of ¨ gas-liquid mass transfer phenomena in microtiter plates," Biotechnology and Bioengineering, vol. 81, no. 2, pp. 178–186, 2003.

27. S. Suresh, V. C. Srivastava, and I. M. Mishra, "Techniques for oxygen transfer measurement in bioreactors: a review," Journal of Chemical Technology and Biotechnology, vol. 84, pp. 1091–1103, 2009.

28. D. A. V. Marques, B. R. Torres, A. L. F. Porto, A. PessoaJunior, and A. Converti, "Comparison of oxygen mass transfer ´ coefficient in simple and extractive fermentation systems," Biochemical Engineering Journal, vol. 47, no. 1–3, pp. 122–126, 2009.

29. H. Tabesh, G. Amoabediny, N. Salehi-Nik, K. Esfahani, H. Derakhshanfar, and B. Zandieh Doulabi, "Use of computerized simulation of engineering parameters in tissue-engineering bioreactors," European Spine Journal, vol. 19, Article ID 1408, 2010.

30. A. B. Yeatts and J. P. Fisher, "Bone tissue engineering bioreactors: dynamic culture and the influence of shear stress," Bone, vol. 48, no. 2, pp. 171–181, 2011.

31. N. Sakamoto, N. Saito, X. Han, T. Ohashi, and M. Sato, "Effect of spatial gradient in fluid shear stress on morphological changes in endothelial cells in response to flow," Biochemical and Biophysical Research Communications, vol. 395, no. 2, pp. 264– 269, 2010.

32. M. B. Simmers, A. W. Pryor, and B. R. Blackman, "Arterial shear stress regulates endothelial cell-directed migration, polarity, and morphology in confluent monolayers," The American Journal of Physiology, vol. 293, no. 3, pp. H1937–H1946, 2007.

33. R. J. McCoy and F. J. O'Brien, "Influence of shear stress in perfusion bioreactor cultures for the development of threedimensional bone tissue constructs: a review," Tissue Engineering B, vol. 16, no. 6, pp. 587–601, 2010.

34. S. D. Waldman, D. C. Couto, M. D. Grynpas, R. M. Pilliar, and R. A. Kandel, "Multi-axial mechanical stimulation of tissue engineered cartilage: review," European Cells and Materials, vol. 13, pp. 66–73, 2007.

35. R. G. Bacabac, T. H. Smit, J. J. W. A. Van Loon, B. Z. Doulabi, M. Helder, and J. Klein-Nulend, "Bone cell responses to highfrequency vibration stress: does the nucleus oscillate within the cytoplasm?" FASEB Journal, vol. 20, no. 7, pp. 858–864, 2006.

36. G. Lemon, J. R. King, H. M. Byrne, O. E. Jensen, and K. M. Shakesheff, "Mathematical modelling of engineered tissue growth using a multiphase porous flow mixture theory," Journal of Mathematical Biology, vol. 52, no. 5, pp. 571–594, 2006.

37. A. Vatsa, T. H. Smit, and J. Klein-Nulend, "Extracellular NO signalling from a mechanically stimulated osteocyte," Journal of Biomechanics, vol. 40, no. 1, pp. S89–S95, 2007.

38. R. G. Bacabac, D. Mizuno, C. F. Schmidt et al., "Round versus flat: bone cell morphology, elasticity, and mechanosensing," Journal of Biomechanics, vol. 41, pp. 1590–1598, 2008.

39. H. J. Henzler, "Particle stress in bioreactors," Advances in Biochemical Engineering/Biotechnology, vol. 67, pp. 35–82, 2000.

40. B. J. H. Zoro, S. Owen, R. A. L. Drake, and M. Hoare, "The impact of process stress on suspended anchorage-dependent mammalian cells as an indicator of likely challenges for regenerative medicines," Biotechnology and Bioengineering, vol. 99, no. 2, pp. 468–474, 2008.

41. V. Bayati, Y. Sadeghi, M. A. Shokrgozar et al., "The evaluation of cyclic uniaxial strain on myogenic differentiation of adiposederived stem cells," Tissue and Cell, vol. 43, no. 6, pp. 359–366, 2011.

42. J. Hatami, M. Tafazzoli-Shadpour, N. Haghighipour, and M. A. Shokrgozar, "Evaluation of Effects of cyclic loading on structural properties of cultured endothelial cell," Modares Journal of Medical Sciences, vol. 12, no. 4, pp. 19–30, 2010.

43. N. Haghighipour, M. Tafazzoli-Shadpour, M. A. Shokrgozar, S. Amini, A. Amanzadeh, and M. T. Khorasani, "Topological remodeling of cultured endothelial cells by characterized cyclic strains," MCB Molecular and Cellular Biomechanics, vol. 4, no. 4, pp. 189–199, 2007.

44. N. Haghighipour, M. Tafazzoli-Shadpour, and A. Avolio, "Residual stress distribution in a lamellar model of the arterial wall," Journal of Medical Engineering and Technology, vol. 34, no. 7-8, pp. 422–428, 2010.

45. N. Haghighipour, M. Tafazzoli-Shadpour, M. A. Shokrgozar, and S. Amini, "Effects of cyclic stretch waveform on endothelial cell morphology using fractal analysis," Artificial Organs, vol. 34, no. 6, pp. 481–490, 2010.

46. F. Safshekan, M. Tafazzoli Shadpour, M. A. Shokrgozar, N. Haghighipour, R. Mahdian, and A. Hemmati, "Intermittent hydrostatic pressure enhances growth factor-induced chondroinduction of human adipose-derived mesenchymal stem cells," Artificial Organs, vol. 36, no. 12, pp. 1065–1071, 2012. BioMed Research International 13

47. A. D. Bakker, K. Soejima, J. Klein-Nulend, and E. H. Burger, "The production of nitric oxide and prostaglandin E2 by primary bone cells is shear stress dependent," Journal of Biomechanics, vol. 34, no. 5, pp. 671–677, 2001.

48. H. . Kaur, R. Carriveau, and B. Mutus, "A simple parallel plate flow chamber to study effects of shear stress on endothelial cells," The American Journal of Biomedical Sciences, vol. 4, no. 1, pp. 70–78, 2012.

49. H. Kang, Y. Fan, and X. Deng, "Vascular smooth muscle cell glycocalyx modulates shear-induced proliferation, migration, and NO production responses," The American Journal of Physiology, vol. 300, no. 1, pp. H76–H83, 2011.

50. B. C. Isenberg, C. Williams, and R. T. Tranquillo, "Smalldiameter artificial arteries engineered in vitro," Circulation Research, vol. 98, no. 1, pp. 25–35, 2006.

51. S. E. Diamantouros, L. G. Hurtado-Aguilar, T. Schmitz-Rode, P. Mela, and S. Jockenhoevel, "Pulsatile perfusion bioreactor system for durability testing and compliance estimation of tissue engineered vascular grafts," Annals of Biomedical Engineering, 2013.

52. M. S. Hahn, M. K. McHale, E. Wang, R. H. Schmedlen, and J. L. West, "Physiologic pulsatile flow bioreactor conditioning of poly(ethylene glycol)-based tissue engineered vascular grafts," Annals of Biomedical Engineering, vol. 35, no. 2, pp. 190–200, 2007.

53. M. T. Zaucha, J. Raykin, W. Wan et al., "A novel cylindrical biaxial computer-controlled bioreactor and biomechanical testing device for vascular tissue engineering," Tissue Engineering A, vol. 15, no. 11, pp. 3331–3340, 2009.

54. N. Haghighipour, S. Heidarian, M. A. Shokrgozar, and N. Amirizadeh, "Differential effects of cyclic uniaxial stretch on human mesenchymal stem cell into skeletal muscle cell," Cell Biology International, vol. 36, no. 7, pp. 669–675, 2012.

55. M. Petrovic, D. Mitrakovi´c, B. Bugarski, D. Vonwil, I. Martin, ´ and B. Obradovic, "A novel bioreactor with mechanical stim- ´ ulation for skeletal tissue engineering," Chemical Industry and Chemical Engineering Quarterly, vol. 15, no. 1, pp. 41–44, 2009.

56. N. Tandon, A. Marsano, R. Maidhof et al., "Surface-patterned electrode bioreactor for electrical stimulation," Lab on a Chip, vol. 10, no. 6, pp. 692–700, 2010.

57. N. Tandon, A. Marsano, C. Cannizzaro, J. Voldman, and G. Vunjak-Novakovic, "Design of electrical stimulation bioreactors for cardiac tissue engineering," Proceedings of the Annual International Conference of the IEEE Engineering in Medicine and Biology Society, vol. 2008, pp. 3594–3597, 2008.

58. M. Radisic, H. Park, H. Shing et al., "Functional assembly of engineered myocardium by electrical stimulation of cardiac myocytes cultured on scaffolds," Proceedings of the National Academy of Sciences of the United States of America, vol. 101, no. 52, pp. 18129–18134, 2004.

59. N. Tandon, A. Marsano, R. Maidhof, L. Wan, H. Park, and G. Vunjak-Novakovic, "Optimization of electrical stimulation parameters for cardiac tissue engineering," Journal of Tissue Engineering and Regenerative Medicine, vol. 5, no. 6, pp. e115– e125, 2011.

60. I. C. Liao, J. B. Liu, N. Bursac, and K. W. Leong, "Effect of electromechanical stimulation on the maturationofmyotubes on aligned electrospun fibers," Cellular and Molecular Bioengineering, vol. 1, pp. 133–145, 2008.

61. J. Malda, T. B. F. Woodfield, F. van der Vloodt et al., "The effect of PEGT/PBT scaffold architecture on the composition of tissue engineered cartilage," Biomaterials, vol. 26, no. 1, pp. 63–72, 2005.

62. A. Fernandes-Platzgummer, M. M. Diogo, R. P. Baptista, C. L. D. Silva, and J. M. S. Cabral, "Scale-up of mouse embryonic stem cell expansion in stirred bioreactors," Biotechnology Progress, vol. 27, no. 5, pp. 1421–1432, 2011.

63. S. Partap, N. A. Plunkett, and F. J. O' Brien, "Bioreactors in tissue engineering," in Tissue Engineering, D. Eberli, Ed., pp. 323–337, 2010.

64. A. B. Yeatts and J. P. Fisher, "Bone tissue engineering bioreactors: dynamic culture and the influence of shear stress," Bone, vol. 48, no. 2, pp. 171–181, 2011.

65. E. Oragui, M. Nannaparaju, and W. S. Khan, "The role of bioreactors in tissue engineering for musculoskeletal applications," The Open Orthopaedics Journal, vol. 5, pp. 267–270, 2011.

66. X. Zhang, C.-A. Burki, M. Stettler et al., "Efficient oxygen ⁚ transfer by surface aeration in shaken cylindrical containers for mammalian cell cultivation at volumetric scales up to 1000 L," Biochemical Engineering Journal, vol. 45, no. 1, pp. 41–47, 2009.

67. L. A. Belfiore, W. Bonani, M. Leoni, and C. J. Belfiore, "Pressuresensitive nutrient consumption via dynamic normal stress in rotational bioreactors," Biophysical Chemistry, vol. 140, no. 1–3, pp. 99–107, 2009.

68. D. Nesic, R. Whiteside, M. Brittberg, D. Wendt, I. Martin, and P. Mainil-Varlet, "Cartilage tissue engineering for degenerative joint disease,"Advanced Drug Delivery Reviews, vol. 58, no. 2, pp. 300–322, 2006.

69. S. R. Khetani and S. N. Bhatia, "Engineering tissues for in vitro applications," Current Opinion in Biotechnology, vol. 17, no. 5, pp. 524–531, 2006.

70. X. Yu, E. A. Botchwey, E. M. Levine, S. R. Pollack, and C. T. Laurencin, "Bioreactor-based bone tissue engineering: the influence of dynamic flow on osteoblast phenotypic expression and matrix mineralization," Proceedings of

the National Academy of Sciences of the United States of America, vol. 101, no. 31, pp. 11203– 11208, 2004.

71. Z.-Y. Zhang, S. H. Teoh, W.-S. Chong et al., "A biaxial rotating bioreactor for the culture of fetal mesenchymal stem cells for bone tissue engineering," Biomaterials, vol. 30, no. 14, pp. 2694– 2704, 2009.

72. S. Lotter and J. Buchs, "Utilization of specific power input ¨ measurements for optimization of culture conditions in shaking flasks," Biochemical Engineering Journal, vol. 17, no. 3, pp. 195– 203, 2004.

73. T. Anderlei, W. Zang, M. Papaspyrou, and J. Buchs, "Online res- ¨ piration activity measurement (OTR, CTR, RQ) in shake flasks," Biochemical Engineering Journal, vol. 17, no. 3, pp. 187–194, 2004.

74. S. A. Freyer, M. Konig, and A. K ¨ unkel, "Validating shaking ¨ flasks as representative screening systems," Biochemical Engineering Journal, vol. 17, no. 3, pp. 169–173, 2004.

75. A. Akgun, C. M ¨ uller, R. Engmann, and J. B ¨ uchs, "Application ¨ of an improved continuous parallel shaken bioreactor system for three microbial model systems," Bioprocess and Biosystems Engineering, vol. 31, no. 3, pp. 193–205, 2008.

76. M. Jamnongwong, K. Loubiere, N. Dietrich, and G. Hebrard, ´ "Experimental study of oxygen diffusion coefficients in clean water containing salt, glucose or surfactant: consequences on the liquid-side mass transfer coefficients," Chemical Engineering Journal, vol. 165, no. 3, pp. 758–768, 2010.

77. J. J. Bellucci and K. H. Hamaker, "Evaluation of oxygen transfer rates in stirred-tank bioreactors for clinical manufacturing," Biotechnology Progress, vol. 27, no. 2, pp. 368–376, 2011. 14 BioMed Research International

78. W. A. Duetz and B. Witholt, "Oxygen transfer by orbital shaking of square vessels and deepwell microtiter plates of various dimensions," Biochemical Engineering Journal, vol. 17, no. 3, pp. 181–185, 2004.

79. S. D. Doig, S. C. R. Pickering, G. J. Lye, and F. Baganz, "Modelling surface aeration rates in shaken microtitre plates using dimensionless groups," Chemical Engineering Science, vol. 60, no. 10, pp. 2741–2750, 2005.

80. P. Therning and A. Rasmuson, "Mass transfer measurements in a non-isothermal bubble column using the uncatalyzed oxidation of sulphite to sulphate," Chemical Engineering Journal, vol. 116, no. 2, pp. 97–103, 2006.

81. D. Cascaval, A.-I. Galaction, E. Folescu, and M. Turnea, "Comparative study on the effects of n-dodecane addition on oxygen transfer in stirred bioreactors for simulated, bacterial and yeasts broths," Biochemical Engineering Journal, vol. 31, no. 1, pp. 56– 66, 2006.

82. M. S. Puthli, V. K. Rathod, and A. B. Pandit, "Gas-liquid mass transfer studies with triple impeller system on a laboratory scale bioreactor," Biochemical Engineering Journal, vol. 23, no. 1, pp. 25–30, 2005.

83. J. M. Seletzky, U. Noack, J. Fricke, S. Hahn, and J. Buchs, ¨ "Metabolic activity of Corynebacterium glutamicum grown on L-lactic acid under stress," Applied Microbiology and Biotechnology, vol. 72, no. 6, pp. 1297–1307, 2006.

84. G. Amoabediny, M. P. H. Abbas, and J. Buchs, "Determination ¨ of CO2 sensitivity of micro-organisms in shaken bioreactors. II. Novel online monitoring method," Biotechnology and Applied Biochemistry, vol. 57, no. 4, pp. 167–175, 2010.

85. G. Amoabediny and J. Buchs, "Determination of CO ¨ 2 sensitivity of micro-organisms in shaken bioreactors. I. Novel method based on the resistance of sterile closure," Biotechnology and Applied Biochemistry, vol. 57, no. 4, pp. 157–166, 2010.

86. J. M. Seletzky, U. Noack, S. Hahn, A. Knoll, G. Amoabediny, and J. Buchs, "An experimental comparison of respiration mea- ¨ suring techniques in fermenters and shake flasks: exhaust gas analyzer vs. RAMOS device vs. respirometer," Journal of Industrial Microbiology and Biotechnology, vol. 34, no. 2, pp. 123–130, 2007.

87. C. Pena, C. P. Peter, J. B ̃ uchs, and E. Galindo, "Evolution ̈ of the specific power consumption and oxygen transfer rate in alginate-producing cultures of Azotobacter vinelandii conducted in shake flasks," Biochemical Engineering Journal, vol. 36, no. 2, pp. 73–80, 2007.

88. M. Scheidle, J. Klinger, and J. Buchs, "Combination of on- ̈ line pH and oxygen transfer rate measurement in shake flasks by fiber optical technique and respiration activity monitoring system (RAMOS)," Sensors, vol. 7, pp. 3472–3480, 2007.

89. A. R. C. Ortigara, P. Foladori, and G. Andreottola, "Kinetics of heterotrophic biomass and storage mechanism in wetland cores measured by respirometry," Water Science and Technology, vol. 64, no. 2, pp. 409–415, 2011.

90. P. Liovic, I. D. Sutalo, R. Stewart, V. Glattauer, and L. Meagher, ̆ "Fluid flow and stresses on microcarriers in spinner flask bioreactors," in Proceedings of the 9th International Conference on CFD in the Minerals and Process Industries, Melbourne, VIC, Australia, 2012.

91. M. Rossi, R. Lindken, P. Hierck B, and J. Westerweel, "Microfluidic system for the study of mechanical and biochemical response of endothelial cells to flow-induced mechanical stimuli," in Proceedings of the 12th Conference on Miniaturized Systems for Chemistry and Life Sciences, 2008.

92. M. Leong Ch, A. Voorhees, and G. B. Nackman T Wei, "Flow bioreactor design for quantitative measurements over endothelial cells using micro-particle image velocimetry," Review of Scientific Instruments, vol. 84, Article ID 045109, 10 pages, 2013.

93. C. V. Nguyen, J. Carberry, and A. Fouras, "Volumetric correlation PIV to measure particle concentration and velocity of microflows," Experiments in Fluids, vol. 52, no. 3, pp. 636–677, 2011.

94. S. P. A. Higgins, C. R. Samarage, D. M. Paganin, and A. Fouras, "Holographic Correlation Velocimetry," in Proceedings of the

9th international symposium on particle image velocimetry, Kobe, Japan, 2011.

95. M. Z. Ismadi, S. Higgins, C. R. Samarage, D. Paganin, K. Hourigan et al., "Optimisation of a stirred bioreactor through the use of a novel holographic correlation velocimetry flow measurement technique," Plos ONE, vol. 8, no. 6, Article ID e65714, 14 pages, 2013.

96. T. Ooms, W. Koek, and J. Westerweel, "Digital holographic particle image velocimetry: eliminating a sign-ambiguity error and a bias error from the measured particle field displacement," Measurement Science and Technology, vol. 19, no. 7, Article ID 074003, 2008.

97. N. G. Deen, B. H. Hjertager, and T. Solberg, "Comparison of PIV and LDA measurement methods applied to the gas-liquid flow in a bubble column," in Proceedings of the10th International Symposium on Applications of Laser Techniques to Fluid Mechanics, Lisbon, Portugal, 2000.

98. Y. Feng, J. Goree, and B. Liu, "Errors in particle tracking velocimetry with high-speed cameras," Review of Scientific Instruments, vol. 82, no. 5, Article ID 053707, 2011.

99. Z. Chara and V. Matousek, "Comparative study of ADV and LDA measuring techniques," in Proceedings of the 6th International Symposium on Ultrasonic Doppler Methods for Fluid Mechanics and Fluid Engineering.

100. H. Chanson, M. Trevethan, and S. I. Aoki, "Acoustic Doppler velocimetry (ADV) in a small estuarine system," in IIAHR Congress, Republic of Korea, 2005.

101. R. I. Abousleiman and V. I. Sikavitsas, "Bioreactors for tissues of the musculoskeletal system," Advances in Experimental Medicine and Biology, vol. 585, pp. 243–259, 2006.

102. E. Cimetta, M. Flaibani, M. Mella et al., "Enhancement of viability of muscle precursor cells on 3D scaffold in a perfusion bioreactor," International Journal of Artificial Organs, vol. 30, no. 5, pp. 415–428, 2007.

103. S. S. Kim, R. Penkala, and P. Abrahimi, "A perfusion bioreactor for intestinal tissue engineering," Journal of Surgical Research, vol. 142, no. 2, pp. 327–331, 2007.

104. V. I. Sikavitsas, G. N. Bancroft, J. J. Lemoine, M. A. K. Liebschner, M. Dauner, and A. G. Mikos, "Flow perfusion enhances the calcified matrix deposition of marrow stromal cells in biodegradable nonwoven fiber mesh scaffolds," Annals of Biomedical Engineering, vol. 33, no. 1, pp. 63–70, 2005.

105. M. Lovett, D. Rockwood, A. Baryshyan, and D. L. Kaplan, "Simple modular bioreactors for tissue engineering: a system for characterization of oxygen gradients, human mesenchymal stem cell differentiation, and prevascularization," Tissue Engineering C, vol. 16, no. 6, pp. 1565–1573, 2010.

106. C. Lee, S. Grad, M. Wimmer, and M. Alini, "The influence of mechanical stimuli on articular cartilage tissue engineering," in Topics in Tissue Engineering, N. Ashammakhi and R. L. Reis, Eds., vol. 2, pp. 1–32, 2005. BioMed Research International 15

107. J. Cerulli, Perfusion Bioreactor for the Development of TissueEngineered Blood Vessels

108. Bachelor thesis., Worcester polytechnic Institute, 2011.

109. M. Radisic, A. Marsano, R. Maidhof, Y. Wang, and G. VunjakNovakovic, "Cardiac tissue engineering using perfusion bioreactor systems," Nature Protocols, vol. 3, no. 4, pp. 719–738, 2008.

110. U. Meyer, A. Buchter, N. Nazer, and H. P. Wiesmann, "Design ¨ and performance of a bioreactor system for mechanically promoted three-dimensional tissue engineering," British Journal of Oral and Maxillofacial Surgery, vol. 44, no. 2, pp. 134–140, 2006.

111. C. A. Chung, C. P. Chen, T. H. Lin, and C. S. Tseng, "A compact computational model for cell construct development in perfusion culture," Biotechnology and Bioengineering, vol. 99, no. 6, pp. 1535–1541, 2008.

112. C. A. Chung, C. W. Yang, and C. W. Chen, "Analysis of cell growth and diffusion in a scaffold for cartilage tissue engineering," Biotechnology and Bioengineering, vol. 94, no. 6, pp. 1138–1146, 2006.

113. M. C. Lewis, B. D. MacArthur, J. Malda, G. Pettet, and C. P. Please, "Heterogeneous proliferation within engineered cartilaginous tissue: the role of oxygen tension," Biotechnology and Bioengineering, vol. 91, no. 5, pp. 607–615, 2005.

114. J. Malda, J. Rouwkema, D. E. Martens et al., "Oxygen gradients in tissue-engineered PEGT/PBT cartilaginous constructs: measurement and modeling," Biotechnology and Bioengineering, vol. 86, no. 1, pp. 9–18, 2004.

115. X. Yan, D. J. Bergstrom, and X. B. Chen, "Modeling of cell cultures in perfusion bioreactors," IEEE Transactions on Biomedical Engineering, vol. 59, no. 9, pp. 2568–2575, 2012.

116. R. D. O'Dea, S. L. Waters, and H. M. Byrne, "A two-fluid model for tissue growth within a dynamic flow environment," European Journal of Applied Mathematics, vol. 19, no. 6, pp. 607–634, 2008.

117. Y. Shi, "Numerical simulation of global hydro-dynamics in a pulsatile bioreactor for cardiovascular tissue engineering," Journal of Biomechanics, vol. 41, no. 5, pp. 953–959, 2008.

118. P. Sucosky, D. F. Osorio, J. B. Brown, and G. P. Neitzel, "Fluid mechanics of a spinner-flask bioreactor," Biotechnology and Bioengineering, vol. 85, no. 1, pp. 34–46, 2004.

119. M. Shakeel, Continuum modelling of cell growth andnutrient transport in a perfusion bioreactor

120. Ph.D. thesis., University of Nottingham, 2011.

121. B. G. Sengers, C. C. van Donkelaar, C. W. J. Oomens, and F. P. T. Baaijens, "Computational study of culture conditions and nutrient supply in cartilage tissue engineering," Biotechnology Progress, vol. 21, no. 4, pp. 1252–1261, 2005.

122. C. A. Chung, C.W. Chen, C. P. Chen, and C. S. Tseng, "Enhancement of cell growth in tissue-engineering constructs under direct perfusion: modeling and simulation,"

Biotechnology and Bioengineering, vol. 97, no. 6, pp. 1603–1616, 2007.

123. F. Coletti, S. Macchietto, and N. Elvassore, "Mathematical modeling of three-dimensional cell cultures in perfusion bioreactors," Industrial and Engineering Chemistry Research, vol. 45, no. 24, pp. 8158–8169, 2006.

124. R. J. Whittaker, R. Booth, R. Dyson et al., "Mathematical modelling of fibre-enhanced perfusion inside a tissue-engineering bioreactor," Journal of Theoretical Biology, vol. 256, no. 4, pp. 533–546, 2009.

125. G. Amoabediny and J. Buchs, "Modelling and advanced under- ¨ standing of unsteady-state gas transfer in shaking bioreactors," Biotechnology and Applied Biochemistry, vol. 46, no. 1, pp. 57–67, 2007.

126. T. Anderlei, C. Mrotzek, S. Bartsch, G. Amoabediny, C. P. Peter, and J. Buchs, "New method to determine the mass ¨ transfer resistance of sterile closures for shaken bioreactors," Biotechnology and Bioengineering, vol. 98, no. 5, pp. 999–1007, 2007.

127. M. Pisu, N. Lai, A. Cincotti, A. Concas, and G. Cao, "Modeling of engineered cartilage growth in rotating bioreactors," Chemical Engineering Science, vol. 59, no. 22-23, pp. 5035–5040, 2004.

128. N. S. Abdullah and D. B. Das, "Modelling nutrient transport in hollow fibre membrane bioreactor for growing bone tissue with consideration of multi-component interactions," Chemical Engineering Science, vol. 62, no. 21, pp. 5821–5839, 2007.

129. P. Yu, T. S. Lee, Y. Zeng, and H. T. Low, "A 3D analysis of oxygen transfer in a low-cost micro-bioreactor for animal cell suspension culture," Computer Methods and Programs in Biomedicine, vol. 85, no. 1, pp. 59–68, 2007.

130. M. Cioffi, F. Boschetti, M. T. Raimondi, and G. Dubini, "Modeling evaluation of the fluid-dynamic microenvironment in tissue-engineered constructs: a micro-CT based model,"

Biotechnology and Bioengineering, vol. 93, no. 3, pp. 500–510, 2006.

131. D. W. Hutmacher and H. Singh, "Computational fluid dynamics for improved bioreactor design and 3D culture," Trends in Biotechnology, vol. 26, no. 4, pp. 166–172, 2008.

132. B. Porter, R. Zauel, H. Stockman, R. Guldberg, and D. Fyhrie, "3-D computational modeling of media flow through scaffolds in a perfusion bioreactor," Journal of Biomechanics, vol. 38, no. 3, pp. 543–549, 2005.

133. H. Dubey, S. K. Das, and T. Panda, "Numerical simulation of a fully baffled biological reactor: the differential circumferential averaging mixing plane approach," Biotechnology and Bioengineering, vol. 95, no. 4, pp. 754–766, 2006.

134. H. Singh, S. A. Eng, T. T. Lim, and D. W. Hutmacher, "Flow modeling in a novel non-perfusion conical bioreactor," Biotechnology and Bioengineering, vol. 97, no. 5, pp. 1291–1299, 2007.

135. Y. Zeng, T.-S. Lee, P. Yu, P. Roy, and H.-T. Low, "Mass transport and shear stress in a microchannel bioreactor: numerical simulation and dynamic similarity," Journal of Biomechanical Engineering, vol. 128, no. 2, pp. 185–193, 2006.

136. C. Provin, K. Takano, Y. Sakai, T. Fujii, and R. Shirakashi, "A method for the design of 3D scaffolds for high-density cell attachment and determination of optimum perfusion culture conditions," Journal of Biomechanics, vol. 41, no. 7, pp. 1436–1449, 2008.

137. F. Boschetti, M. T. Raimondi, F. Migliavacca, and G. Dubini, "Prediction of the micro-fluid dynamic environment imposed to three-dimensional engineered cell systems in bioreactors," Journal of Biomechanics, vol. 39, no. 3, pp. 418–425, 2006.

138. B. Pouran, G. Amoabediny, S. Saghafinia, and M. P. Haji Abbas, "Characterization of interfacial hydrodynamics in a single cell of shaken microtiter plate bioreactors applying computational fluid dynamics technique," Procedia Engineering, vol. 42, pp. 924–930, 2012.

139. P. Yu, T. S. Lee, Y. Zeng, and H. T. Low, "Fluid dynamics of a micro-bioreactor for tissue engineering," Fluid Dynamics and Materials Processing, vol. 1, pp. 235–246, 2005.

140. B. Bilgen and G. A. Barabino, "Location of scaffolds in bioreactors modulates the hydrodynamic environment experienced by engineered tissues," Biotechnology and Bioengineering, vol. 98, no. 1, pp. 282–294, 2007.

141. H. O. Tabrizi, G. Amoabediny, B. Moshiri et al., "Novel dynamic model for aerated shaking bioreactors," Biotechnology and Applied Biochemistry, vol. 58, no. 2, pp. 128–137, 2011.

Chapter 3

CFD Prediction of the Turbulent Flow Generated in Stirred Square Tank by a Rushton Turbine

W. Chtourou, M. Ammar, Z. Driss, and M. S. Abid

National School of Engineers of Sfax (ENIS), Department of Mechanical Engineering, Laboratory of Electromechanical Systems (LASEM), Sfax, Tunisia

ABSTRACT

The Computational Fluid Dynamics (CFD) have been used in the analysis and design of agitated vessel. Most of the researches done in this area are limited to the baffled or unbaffled stirred tank. In this paper, we have been interested in studying of the new design. Particularly, the flow and turbulence fields in square vessel stirred by a standard Rushton turbine have been simulated by means of CFD techniques. The Navier-Stokes equations governing the phenomenon of transfer of momentum are solved by a discretization method for finite volume. The MRF approaches can be used in simulation of the steady state problem. The numerical results from the application of CFD code Fluent with the stationary approach Multi Reference Frame (MRF) are presented in the planes containing the blade. The validation of CFD results with experimental measurements showed a good agreement.

INTRODUCTION

Mixing is a very common operation in the process industries, usually performed by mechanical agitation.

Stirred tanks for the homogenization of single or several phases are among the most commonly used equipment in the chemical and biochemical processes. The accumulation of data on the operation of industrial agitators on the one hand, and a major research effort done in collaboration with academics, on the other hand, can provide experimental and theoretical support needed to technological advances. In view of the above, in the past, the impeller design and these characteristics have been the goal of most work. The flow generated by the Rushton turbines (RT) has been subjected to detailed experimental and computational studies. In the literature, several works relating to the study of these types of turbines are already published. As a guide, one can cite the experimental work conducted by Rushton [1] who studied the effect of a Rushton turbine on the evolution of the power number. These studies have shown that the design plays a key role in determining power consumption and agitator efficiency. It is worth noting that the tank design plays a key role in determining the power consumption and the agitator efficiency. Although this has become more routine for baffled vessels Montante et al., [2] [3] (2004, 2005) the unbaffled heavily swirling case is less well characterised by Armenante et al., [5] (1994), although a recent advanced LES study is noteworthy Alcamo et al., [6] . Even so, until very recently CFD only provided information on fluid mechanics, although just recently it is being exploited to predict mixing curves involving the evolution of concentration fields Yeoh et al. [7] , Montante et al. [3] . Yet, there are cases in which the use of unbaffled tanks may be desirable. First, baffles are usually omitted in the case of very viscous fluids (Re < 20), where they, giving rise to dead zones, would actually worsen the mixer performance, and where vortex formation is inhibited by the low rotational speed and by the high friction on the cylindrical wall Nagata [8] Unbaffled tanks are also advisable in crystallisers, where the presence of baffles may promote the particle attrition phenomenon Mazzarotta et al. [9] and Derkson, J. [4] . Finally, unbaffled tanks give rise to higher fluid-particle mass transfer rates for a given power consumption, which may be desirable in a number of processes Bakker et al. [10]

Before this important flow of information in very little mechanical agitation is the work that has treated the influence of the square tank geometry. Among these works J. Kilander et al. [11] , studied experimentally the hydrodynamic parameters of turbulence in a square tank stirred with an axial impeller type A310 using the approach (PIV) Large Eddy Particle Image Velocimetry. Suzanne M. Kresta et al. [12] conducted an experimental study of the batch mixture into a square tank stirred by using a spectrophotometer to a number of turbine type. This study shows that the correlation Grenville for determining the mixing time is checked for a square tank. J. Kilander et al.[13] determined the spatial and temporal evolution of the floc size distribution in a square 7.3 L tank stirred with an A310 hydro foil impeller investigated using PIV and image analysis. J. Kilander et al. [14] studied the size effect of the square tank on the mechanism of flocculation. The results show that there are large spatial differences in the mean size and the shape of the floc size distribution within the tanks but that the differences decrease as tank size increases from 5 to 28 l. However, increasing the size further to 560 increases the spatial variation.

View the lack of numerical results to evaluate and quantify the agitated vessel in square tank driven by a Rushton turbine, we are interested in this paper to the numerical simulation this type of system in fully turbulent regime.

NUMERICAL APPROACH

Discretization Scheme

Some authors have investigated the effect of the discretization scheme on the accuracy of the predicted flow and in most cases have shown that the choice of discretization scheme has little or no effect on the solution. Bucato et al. [15] compared a hybrid scheme (upwind-central differencing) and the high-order QUICK scheme and found that predicted mean velocities did not differ appreciably. These authors concluded that for their model of 97,000 control volumes, numerical diffusion effects associated with the first-order upwind discretization scheme were not significant and that turbulent diffusion was largely dominant. However, no comparison of the effect of discretization on turbulence quantities was presented and it is not clear whether the results were grid independent. Aubin et al. [16] investigated three discretization methods (first-order upwind, upwind-central hybrid, QUICK) on a grid of 155,000 control volumes and found that the choice of the discretization scheme had no effect on the mean velocities, except that the upwind scheme was found to under-predict the swirling region below the impeller. However, all three discretization schemes were found to under-predict the turbulent kinetic energy, this being most severe for the first-order upwind scheme in the impeller discharge stream.

Impeller Rotation Modelling

Modelling the impeller rotation is complex as the relative motion between the rotating impeller blades and the stationary tank wall causes a cyclic variation of the solution domain. Approaches to modelling the impeller rotation include using experimentally determined impeller boundary conditions, specifying momentum source/sink terms on the impeller blades or incorporating rotating and stationary reference frames. Two commonly used models are the Multiple Reference Frames and Sliding Mesh models. In these models, the solution domain is divided into an inner region containing the rotating impeller and an outer region containing the stationary tank wall. For the MRF model, steady-state calculations are performed with a rotating reference frame in the impeller region and a stationary reference frame in the outer region. In this way, the effects of the blade rotation are accounted for by virtue of the frame of reference, allowing for explicit modelling of the impeller geometry. For the SM model, the impeller region is allowed to slide relative to the outer region in discrete time steps and timedependent calculations are performed using implicit or explicit interpolation of data at successive time-steps. Being time dependent, the SM method is the more accurate representation of the actual phenomenon of the impeller rotation but is computationally demanding.

Reynolds Stress Modeling

There have been two attempts partially focused on the RSM for the flow generated by Rushton turbine in a baffled stirred vessel. First one, Bakker and Van den Akker [17] employed the simplified RSM, i.e. algebraic stress model (ASM) using the IBC method to model the flows produced by a Rushton turbine. Their objective was to improve the predictive capabilities of CFD modeling by accounting anisotropy using less computational intensive ASM model (simplified RSM). Their study concluded that the results predicted by the ASM compare better with the experimental data than those predicted by the standard k – ε model. Oshinowo et al. [18] performed the CFD study using different turbulence models like, k – ε, RNG k – ε and RSM for the prediction of tangential velocity distribution in a baffled vessel using multiple reference frame (MRF) model. The tangential velocity distribution above the impeller has been correctly predicted. They attributed the occurrence of the counter-intuitive reverse swirl in the simulation to poor convergence and coarse grid density. It may be pointed out, however, that both the aforesaid investigations have shown comparisons in only a small region while most of the vessel region remained unexplored. More importantly, both the studies have not presented the predictive capability of CFD models for the turbulent kinetic energy and the turbulent energy dissipation rate.

STIRRED VESSEL CONFIGURATION

The tank geometry employed in this work is a square vessel depicted in Figure 1. The tank length T of 0.19 m and was stirred by a Rushton turbine. The number of blade was six. The disk mounted impeller 0.095 m in diameter (D/T = 0.5), with standard proportions between parts (a = D/4, b = D/5, d = 3/4D. The impeller is installed along the axis of the tank with a shaft diameter of d_s = 0.001 m and an off-bottom clearance of C = T/3. The vessel was 0.19 m tall (H = T) and was provided with a flat lid that inhibited the central vortex formation as well as any surface aeration. The tank was filled with pure water at room temperature. The rotational speed of the impeller N was 200 rpm (3.33 rps), resulting in a Reynolds number of about 4.2×10^4. In the numerical simulations, the fluid was assumed to be incompressible, isothermal and Newtonian with a density of 1000 kg·m^{-3} and a dynamic viscosity of 1×10^{-3} Ns·m^{-2}.

NUMERICAL MODELS

The CFD code Fluent with the multiple reference frames (MRF) was used for direct modeling of the impellertank interaction. Thus, no LDA data was used to obtain the CFD solution of the differential equations for momentum transfer.

Boundary Conditions

The computational domain was split into two cylindrical zones. One of which is assumed to rotate with the impeller angular velocity of $\omega = 2\varpi N$, while the remaining space can be modeled with a stationary reference. In that zone, the options of centrifugal and Coriolis forces were activated. The outer zone was stationary relative to the tank walls. Because the symmetry of the configuration, the half of the area will be analyzed. Figure 2 shows the subdivision of the computational domain with the MRF approach. For the closure of the above equations different turbulence models have been employed and these are described in the following section. The numerical solution of these equations was achieved by a finite-volume method (Burns & Wilkes, 1987), together with the Rhie and Chow (1983) algorithm to prevent "chequer-board" oscillations, both implemented in the computer code Fluent. Because there is no explicit equation for the pressure, special techniques have been devised to extract it in an alternative manner. The most well known of these techniques is the SIMPLE algorithm, or Semilmplicit Method for Pressure-Linked Equations, adopted by Patankar [19] to couple the continuity and NavierStokes equations. The hybrid-upwind discretisation scheme was used for the convective terms. The simulation was also performed using the standard differencing scheme in conjunction with the MRF method to test whether the use of a different scheme would have any impact on the solution. The computed flow field is identical to that obtained with the hybrid scheme and therefore the differencing scheme had no effect on the predictions. In all cases, conventional linear-logarithmic called wall functions were used on solid walls, while the free surface was treated as a symmetry plane. The solution convergence was carefully checked by monitoring the residuals of all variables as well as physical values of the swirl velocity. Residuals were dropped to the order of 10 - 5 or less, which is at least one order of magnitude tighter than Fluent's default criteria.

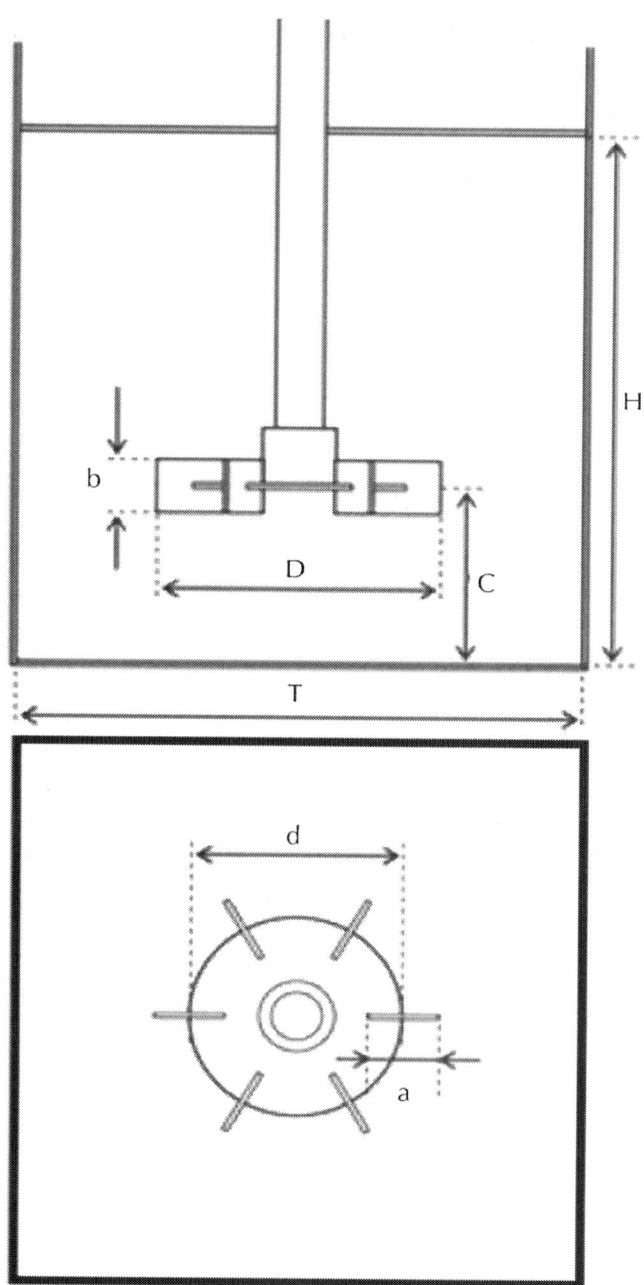

Figure 1: Vessel geometry.

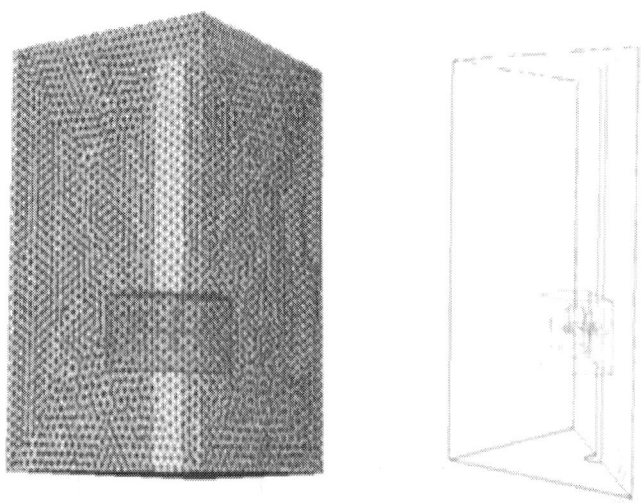

Figure 2: Grid generation and computational domain.

Meshing

The grid elements can affect the quality of the solution. In fact, a preliminary grid convergence study was carried out in order to verify that the solution is a grid independent. For different types of meshes, we are interested to compare the numerical results with experimental data. The number of grid nodes in both zones, inner and outer mesh was systematically increased throughout the vessel. Generally, the numerical simulation with a coarse grid has a large deviation from the reference results. Gradually refining the grid, we note that the appearance of results stabilizes and approaches to the reference values. Also, it's noted that it is unnecessary to further refine the mesh, because the calculation time becomes very large and the results found are almost the same. Thus, we show that there is a compromise between the mesh refinement and choice of computing time. In this work, we have used regular grid that containing hexahedral elements for the full domain simulation in order to conserve flux in each cell and to resolve the steep gradients in the impeller region. The total grid size of 115,000 was used for the full tank simulation. Impeller region (less than 5.5% of tank volume) was meshed with 15% of the total grid size used for the full tank in order to resolve the steep gradients in the impeller region. Similarly, near vessel wall were meshed with the dense grid. In this zone, corresponding to the fully-turbulent layer, the Y+ values is equal to 60.

Governing Equations

By all the computational approaches used in this work, the equations to be solved are the continuity and momentum equations. The continuity equation is a statement of conservation of mass. For a constant density fluid, it takes the form:

$$\operatorname{div} \boldsymbol{v} = 0$$

(1)

The momentum equations are a statement of conservation of momentum in each of the three components. The three momentum equations are collectively called the Navier-Stokes equations. In addition to momentum transport by convection and diffusion, several momentum sources are also involved. In cylindrical coordinates (r, , z), the momentum equations take the form:

$$\operatorname{div}\left(\rho u u_r\right) - \rho \frac{u_r^2}{r} = -\frac{\partial P}{\partial r} + \operatorname{div}\left(\tau_r\right) - \frac{\tau_{\theta\theta}}{r} + F_r$$

(2)

$$\operatorname{div}\left(\rho u u_\theta\right) + \rho \frac{u_r u_\theta}{r} = -\frac{1}{r}\frac{\partial P}{\partial \theta} + \operatorname{div}\left(\tau_\theta\right) - \frac{\tau_{r\theta}}{r} + F_\theta$$

(3)

$$\operatorname{div}\left(\rho u u_z\right) = -\frac{\partial P}{\partial z} + \operatorname{div}\left(\tau_z\right) + F_z$$

(4)

The total pressure P is defined by the following equation:

$$P = P_{stat} + \frac{2}{3}\rho k$$

(5)

$\bar{\bar{\sigma}}$ is the viscous stress tensor

$$\bar{\bar{\sigma}} = \begin{pmatrix} \tau_{rr} & \tau_{\theta r} & \tau_{zr} \\ \tau_{r\theta} & \tau_{\theta\theta} & \tau_{z\theta} \\ \tau_{rz} & \tau_{\theta z} & \tau_{zz} \end{pmatrix}$$

(6)

Within:

$$\tau_{rr} = 2\mu_e \frac{\partial u_r}{\partial r}$$

(7)

$$\tau_{\theta\theta} = 2\mu_e \left(\frac{1}{r}\frac{\partial u_\theta}{\partial \theta} + \frac{u_r}{r} \right)$$

(8)

$$\tau_{zz} = 2\mu_e \frac{\partial u_z}{\partial z}$$

(9)

$$\tau_{r\theta} = \mu_e \left(\frac{1}{r}\frac{\partial u_r}{\partial \theta} + r\frac{\partial}{\partial r}\frac{u_\theta}{r} \right)$$

(10)

$$\tau_{rz} = \mu_e \left(\frac{\partial u_r}{\partial z} + \frac{\partial u_z}{\partial r} \right)$$

(11)

$$\tau_{\theta z} = \mu_e \left(\frac{1}{r}\frac{\partial u_z}{\partial \theta} + \frac{\partial u_\theta}{\partial z} \right)$$

(12)

F_r, F_θ and F_z are respectively the centrifugal force, the Coriolis terms and the gravity. The expressions of these three terms are given in the following form:

$$F_r = \rho\left(\omega^2 r + 2\omega u_\theta\right)$$

(13)

$$F_\theta = \rho\left(-2\omega u_r\right)$$

(14)

$$F_z = -\rho g$$

(15)

For the second-order model (RSM) used in this work, the turbulent kinetic energy and its dissipation rate are given by the system of equations as follows:

$$\frac{\partial(\rho k)}{\partial t} + \frac{\partial(\rho u_i k)}{\partial x_i} = \frac{\partial}{\partial x_j}\left[\left(\mu + \frac{\mu_t}{\sigma_k}\right)\frac{\partial k}{\partial x_j}\right] + \frac{1}{2}P_{ij} - \rho\varepsilon$$

(16)

$$\frac{\partial(\rho\varepsilon)}{\partial t} + \frac{\partial(\rho u_i \varepsilon)}{\partial x_i} = \frac{\partial}{\partial x_j}\left[\left(\mu + \frac{\mu_t}{\sigma_\varepsilon}\right)\frac{\partial\varepsilon}{\partial x_j}\right] + \frac{1}{2}\frac{\varepsilon}{k}C_{1\varepsilon}P_{ij} - C_{2\varepsilon}$$

(17)

Due to gradients ways, the RSM model is characterized by the production term of the turbulent kinetic energy defined as follow:

$$P_{ij} = -\rho\left[\overline{u_i'u_k'}\frac{\partial u_j}{\partial x_k} + \overline{u_j'u_k'}\frac{\partial u_i}{\partial x_k}\right]$$

(18)

This module requires the use of empirical constants given in Table 1 as follows.

RESULTS

Distribution of the Average Velocity in the Plane r-θ

Figure 3 shows the spatial distribution of mean velocity in three planes of the tank. These planes were located symmetrically above and below the plane containing the stirrer. In the central plane containing the turbine Figure 3(b), the seat of the maximum average speed is localized in the area swept by the turbine and decreases between the blade and at the turbine disk. In the area between the tips of turbine blades and the walls of the tank, average speed is almost constant. At the walls of the tank, we note the zero values of average speed. In addition, the average speed is dropped in passing clockwise through the corner of the field. In conclusion, the square geometry of the tank destroys the tangential component of velocity and prevents the generation of vortices in the vicinity of the turbine. For the other two plans in Figure 3(a) and Figure 3(c) the spatial distribution of the average velocity is very different than the previous plan. The area around the axis of the stirring spindle is characterized by a relatively constant average speed and exchanges in this field are purely convective. Near the walls of the tank we have a nonzero mean velocity. Hence the existence of a lateral flow parallel to the walls that follows the formation of recirculation zones at the corners of the tank.

Table 1: Constant for the rms model

Constants	$C_{1\varepsilon}$	$C_{1\varepsilon}$	σk	$\sigma \varepsilon$
Value	1.44	1.92	0.85	1.0

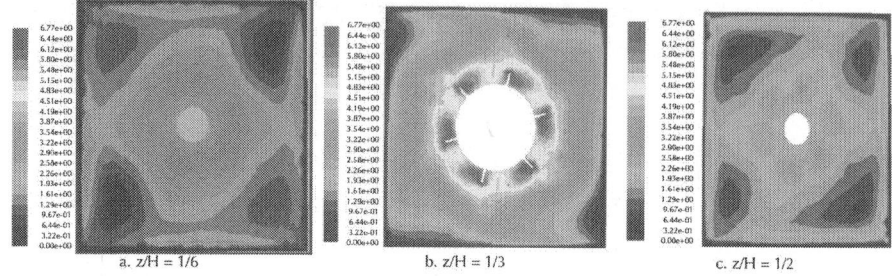

a. z/H = 1/6   b. z/H = 1/3   c. z/H = 1/2

Figure 3: Average velocity in the plane r-θ.

Distribution of the Average Velocity in the Plane r-z

In Figure 4, it is presented the spatial distribution of the average velocity in the plane containing the turbine blade TR6. View the geometry of the tank recovery plans are not the same size. From these results, we note the appearance of a wake maximum value that develops in the area swept by the turbine. The characteristics of radial jet turbine are kept for the three study designs: a large radial jet that develops at the end of the blade and an intense axial flow upward to the underside of the shaft that powers the turbine. View these dimensions and if the central plane Figure 4(a), the radial jet generates a low axial flow at the wall. Also note a recirculation loop that forms in the lower part of the plan. In the case of the plane of Figure 4(b), the spatial distribution of the average speed mark the disappearance of the recirculation loop in the lower part of the tank and the appearance of heavy traffic in the upper part. The radial jet created by the turbine generates an axial flow at the wall. The spatial distribution of mean velocity in the plane of the Figure 4(c) ascended the presence of a plane that intensify the radial jet which reaches the lateral surface and is transformed into two axial jets up and down. The upward axial jet is more intense. It gives rise to a recirculation loop extended well into the upper area of the tank thereby feeding back the turbine.

In Figure 5 is presented the average velocity in the r-z plane located between two blades. Overall, the flow is less intense than in the planes containing the blades. The plan of Figure 5(a) is the plan as a lover of the central plane, where the spatial distribution of the average speed is more intense than that of the plane below the central plane Figure 5(b). Thus the passage of the flow through the central plane destroyed the tangential component of velocity and evidence of slowing the flow downstream of the corner of the square tank. As interpreted, the geometry of the vessel plays a very important role in the destruction of the tangential component of velocity and therefore prevents the formation of vortices around the impeller.

Distribution of Turbulent Kinetic Energy

Figure 6 presents the evolution of the distribution of the turbulent kinetic energy in the square vessel. The three planes in Figure 6 defined respectively by the axial coordinate equal to $z = 1/2$, $z = 1/3$ and $z = 1/6$. Maximum values are recorded in the plane containing the turbine at the axial coordinate $z = 01/3$. These values are located near the flat face of the square tank. The turbulent kinetic energy is very low in the corners of the tank at full plans. For the other two axial positions turbulent kinetic energy falling in a remarkable way. The minimum values are located near the axis and corners of the square tank.

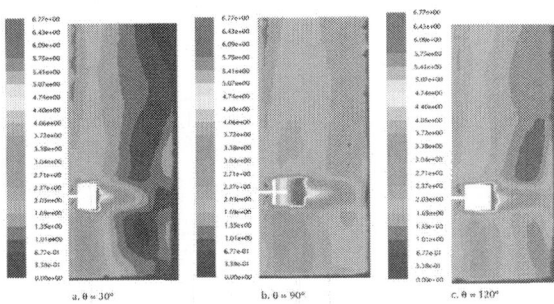

a. θ = 30°   b. θ = 90°   c. θ = 120°

Figure 4: The average velocity in the plane r-z containing the blade.

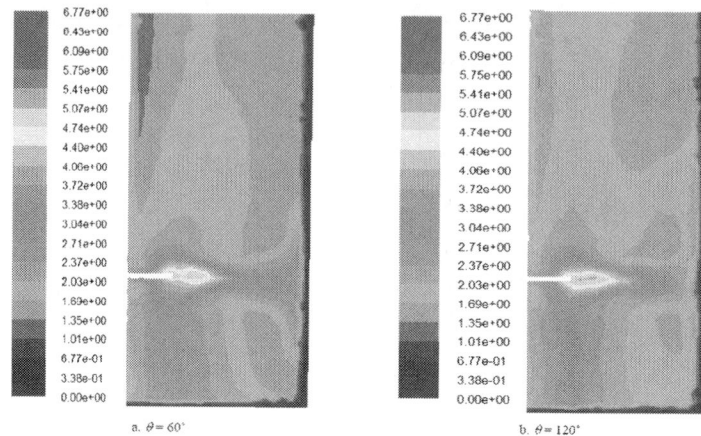

Figure 5: Mean velocity in the plane r-z between blades.

Figure 7 presents the turbulent kinetic energy in the plane containing the blades. The maximum values of turbulent kinetic energy are noted in the plane containing the central blade q = 90°. These values are located near the blade and diffused up the corner of the square tank. There are differences between the plane q = 30° and the plane q = 150°: the thirst is the minimum values of turbulent kinetic energy are noted near the wall in the angular position q = 150°. The second concerned the spatial partition of the turbulent kinetic energy in the proximity of the Blade.

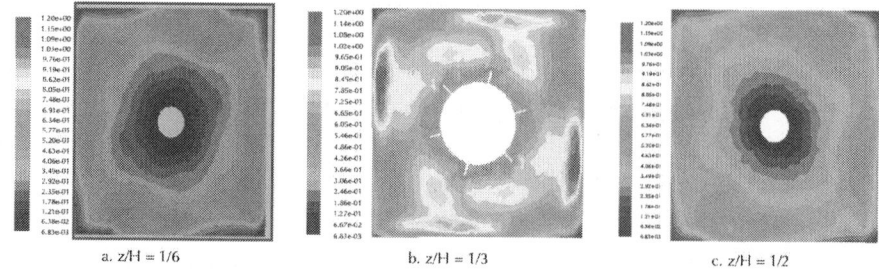

Figure 6: Turbulent kinetic energy in the plane r- .

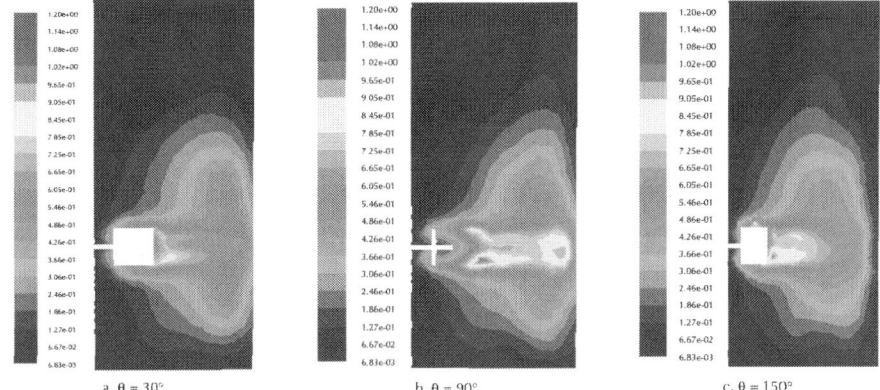

a. θ = 30°   b. θ = 90°   c. θ = 150°

Figure 7: Turbulent kinetic energy in the plane r-z containing the blade.

Figure 8 presented the turbulent kinetic energy in the planes located at the angular position q = 60° and q = 120°. We note that, the tank design and the direction of the turbine rotation affected the partition of the turbulent kinetic energy. For these reasons, the partitions of the turbulent kinetic energy are more intense in the plane q = 120° and the maximum value are located in the proximity of the tank wall.

Distribution of Dissipation Rate of the Turbulent Kinetic Energy

In Figure 9, we presented the distribution spatial of the dissipation rate of the turbulent kinetic energy in planes located respectively at the axial coordinate equal to z = 0.4, z = 0.33 and z = 0.25. Globally, the geometry of the vessel affects the spatial distribution of dissipation rate of the turbulent kinetic energy. We note that the dissipation of energy is in the plane containing the turbine. In this plan the energy is dissipated around the agitation blade and near the walls of the square tank. From both sides of this plan (z/H = 1/6, z/H = 1/2), the dissipation rate of turbulent kinetic energy was near the walls of the square tank.

Figure 10 present the evolution of the distribution of the dissipation rate of the turbulent kinetic energy in planes containing the blade. These results confirm that the dissipation rate of the turbulent kinetic energy was essentially in the region swept by the impeller turbine. We noted, that the maximum of the dissipation was form the angular position q = 90˚.

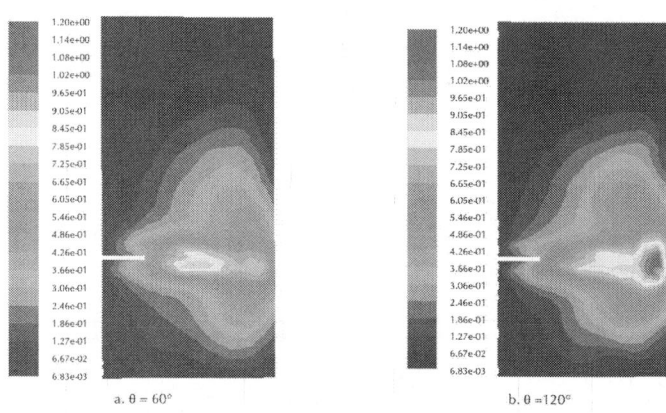

a. θ = 60°   b. θ = 120°

Figure 8: Turbulent kinetic energy in the plane r-z between blades.

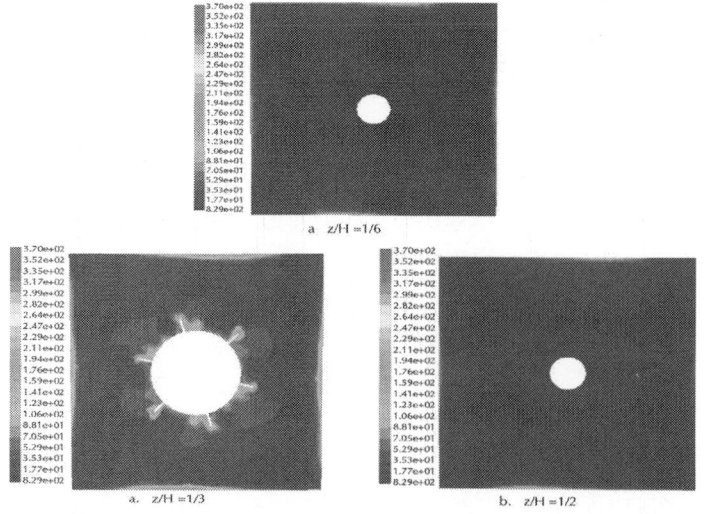

a z/H = 1/6

a. z/H = 1/3   b. z/H = 1/2

Figure 9: Distribution of dissipation rate of the turbulent kinetic energy in the plane r-q.

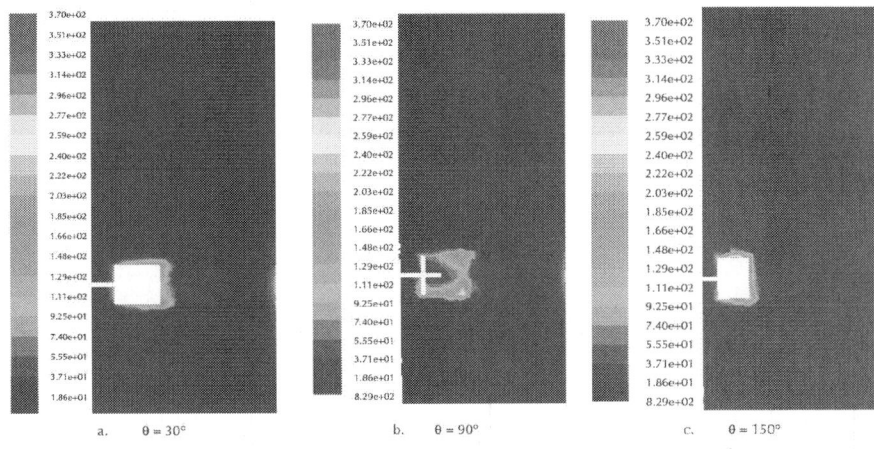

a.   θ = 30°   b.   θ = 90°   c.   θ = 150°

Figure 10: Distribution of dissipation rate of the turbulent kinetic energy in the plane r-z.

Distribution of Turbulent Viscosity

Figure 11 shows the evolution of the turbulent viscosity distribution in the same planes adopted in the previous paragraphs. We noted, the maximum value of the turbulent viscosity are located at the axial position z/H = 1/2. Globally, we find that the turbulent viscosity is very weak on the plane containing the impeller z/H = 1/3. In the symmetrical plane of this plane, it grows a wake characterized by relatively high values. At the proximity in the sidewall of the square tank, the turbulent viscosity drops rapidly.

Figure 12 show the spatial distribution of the turbulent viscosity in three planes containing the impeller blade. The turbulent viscosity has remained rather high in the field swept by blades of the turbine. Very near to the walls and around the turbine, the turbulent viscosity has made a very fast fall. By comparing the three planes between them, it has been noted that the maximal values of the turbulent viscosity in angular position q = 150°. In general, the difference in the spatial distribution of turbulent viscosity in the plans shows the influence of tank geometry on the hydrodynamic structure.

Axial Profiles of the Velocity

Axial Profiles of the Velocity to an Angular Position (= 15°)

Figure 13 presents the axial profiles of the three components of the velocity at the angular position q = 15°. In this position, four radial positions are studied. We note that the radial component of the velocity Figure 13(a) keeps the same appearance for all four radial positions. Except that the intensity of the velocity decreases away from the turbine with a maximum value of U/v_{tip} = 0.3 for a radial position r/R = 0.5 to a value of U/v_{tip} = 0.1 at r/R = 0.95. Figure 13(b) shows the profile of the tangential component of the velocity is established for the radial position r/R = 0.5. Beyond this radial position, we note that the tangential component of the velocity is almost constant. The axial evolution of the axial component of velocity is shown in Figure 13(c). These profiles are symmetrical in appearance: for r/R < 0.68 intense axial jet turbine fuels and for r/R > 0.8 it will reverse the side of the vessel wall.

Axial Profiles of the Velocity to an Angular Position $(\theta = 45°)$

Figure 14 shows the axial profiles of the velocity for four positions. In Figure 14(a), the radial component of the radial velocity for this position is not affected by the geometry of the tank and the maximum values correspond to a radial position r/R = 0.84.

Figure 14(b) presents the tangential component of velocity. This compo- nent is characterized by a value of velocity v/v_{tip} = 0.2 which remains constant on both sides of the turbine. In the area swept by the impeller the tangential component presents the maximum value. This maximum is v/v_{tip} = 0.4 at the radial position r/R = 0.62. Figure 14(c) shows the axial component of velocity. We find that the jet axial turbine that powers extends to a radial position r/R = 0.84. Note that for the radial position r/R = 0.9 the jet is weighty in descending lower zone of the tank but the jet ascending in the upper part is small and it's zero in the axial position between z/H = 0.75 and z/H = 1.

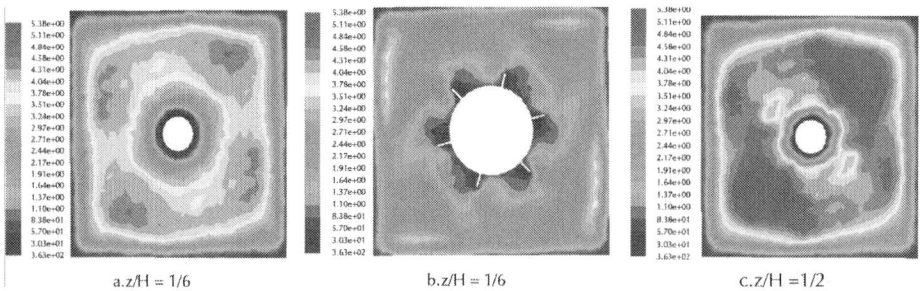

a.z/H = 1/6   b.z/H = 1/6   c.z/H =1/2

Figure 11: Distribution of turbulent viscosity in r θ -q plane.

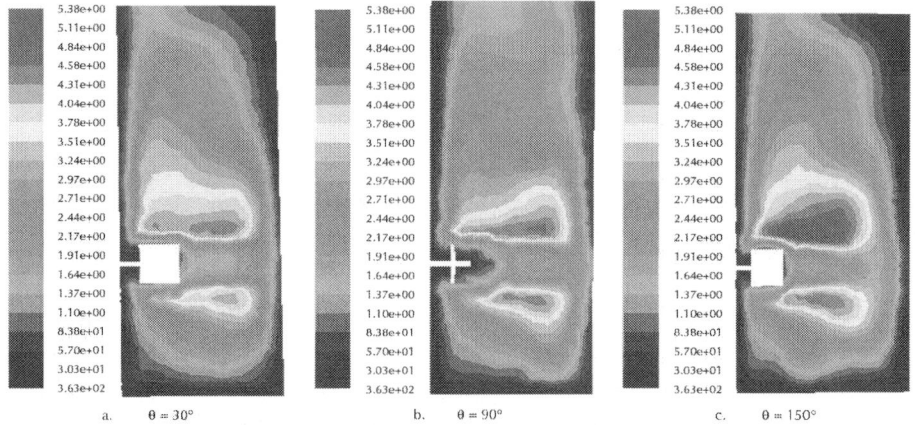

a.   θ = 30°   b.   θ = 90°   c.   θ = 150°

Figure 12: Distribution of turbulent viscosity in r θ -q plane.

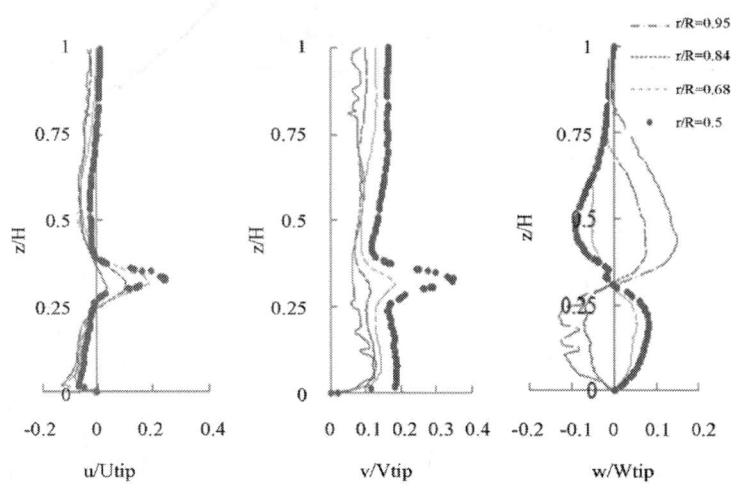

Figure 13: Axial profiles of velocity components at θ = 15°.

Axial Profiles of the Velocity to an Angular Position ($\theta = 75°$)

Figure 15 shows presents the evolution of the axial velocity at the angular position q = 75°. Globally, we noted that the axial component decreases at r/R = 0.35 and r/R = 0.56. This decrease causes an increase of two other components in this region. Near the vessel tank, the axial component of velocity is intense and an axial jet appeared. For against, the two other components of velocity are very low.

COMPARISON WITH EXPERIMENTAL RESULTS

Using the house expérience, an experimental study was conducted to characterize the sturied tank. Specifically, we are interested in determining the variation of power number as a function of Reynolds number in the case of a square tank equipped with a standard Rushton turbine. To verify our computer results, the power number N_p were measured and compared with the number calculated from the CFD code as shown in Figure 16. In this comparison, the average error between the experimental results and the numerical results is equal to 5%. The good agreement confirmed the validity of the computer method.

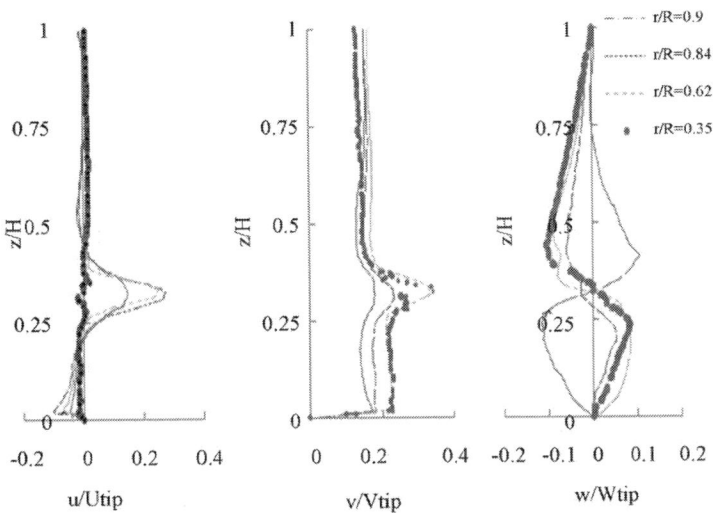

Figure 14: Axial profiles of velocity components at $\theta = 45°$.

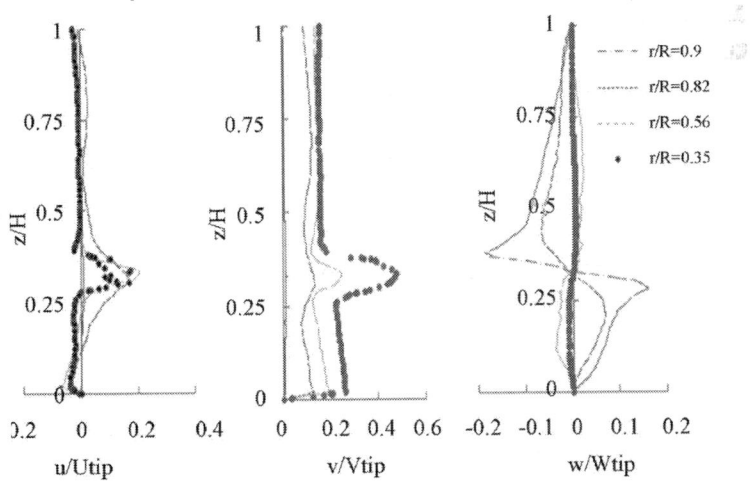

Figure 15: Axial profiles of velocity components at $\theta = 75°$.

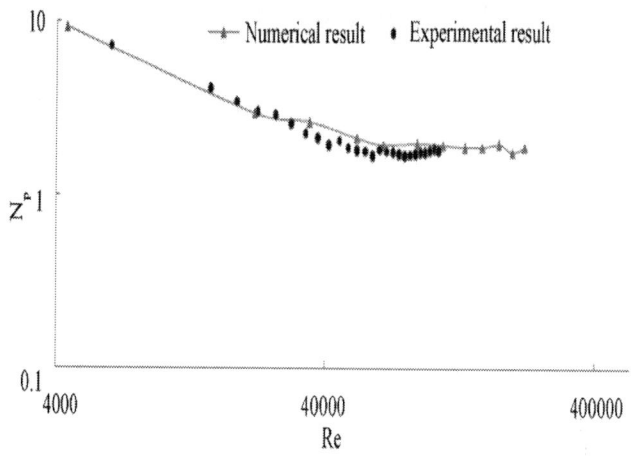

Figure 16: Power number N_p against reynolds number Re of rushton turbine in square tank.

CONCLUSIONS

A numerical study was performed to investigate the hydrodynamic structure generated by a six-blade Rushton turbine in a square vessel. To predict the effect of the tank design on the flow pattern, we have used the MRF approach with the rsm turbulent model. Particularly, we have presented the average velocity and the turbulent kinetic energy characteristics in different planes. To validate the CFD results, the power consumption of a stirred square tank equipped with Rushton impellers was predicted by using both experimental and simulation proceedings. This study needs further investigation both by experimental measurements of the local hydrodynamic with PIV technique and the development of the CFD method with the unsteady sliding mesh approach.

REFERENCES

1.   Rushton, J.H., Costich, E.W. and Everett, H.J. (1950) Power Characteristics of Mixing Impellers. Chemical Engineering Progress, 46, 467-476.

2.   Montante, G., Lee, K.C., Brucato, A. and Yianneskis, M. (2004) Numerical Simulation of the Dependency of Flow Pattern on Impeller Clearance in Stirred Vessels. Chemical Engineering Science, 56, 3751-3770. http://dx.doi.org/10.1016/S0009-2509(01)00089-6

3.   Montante, G., Mostek, M., Jahoda, M. and Magelli, F. (2005) CFD Simulations and Experimental Validation of Homogenisation Curves and Mixing Time in Stirred Newtonian and Pseudoplastic Liquids. Chemical Engineering Science, 60, 2427-2437. http://dx.doi.org/10.1016/j.ces.2004.11.020

4.   Derkson, J. (2002) Confined and Agitated Swirling Flows with Applications in Chemical Engineering. Flow Turbulence and Combustion, 69, 3-33. http://dx.doi.org/10.1023/A:1022419316418

5.   Armenante, P.M. Changgen, L., Chou, C., Fort, I. and Medek, J. (1997) Velocity Profiles in a Closed, Unbaflled Vessel: Comparison between Experimental LDV Data and Numerical CFD Redictions. Chemical Engineering Science, 52, 3483-3492. http://dx.doi.org/10.1016/S0009-2509(97)00150-4

6.   Alcamo, R., Micale, G. Grisafi, F., Brucato, A. and Ciofalo, M. (2005) Large-Eddy Simulation of Turbulent Flow in an Unbaffled Stirred Tank Driven by a Rushton Turbine. Chemical Engineering Science, 60, 2303-2316. http://dx.doi.org/10.1016/j.ces.2004.11.017

7.   Yeoch, S.L., Papadakis, G. and Yianneskis, M. (2005) Determination of Mixing and Degree of Homogeneity in Stirred Vessels with Large Eddy Simulation. Chemical Engineering Science, 60, 2293-2302. http://dx.doi.org/10.1016/j.ces.2004.10.048

8.   Nagata, S. (1975) Mixing Principles and Applications. John Wiley & Sons Halstead Press, Japan.

9.   Mazzarotta, B. (1993) Comminution Phenomena in Stirred Sugar Suspensions. A.I.Ch.E. Symposium Series, 89, 112-117.

10.   Bakker, A., Fasano, J.B. and Myers, K.J. (1994) Effects of Flow Pattern on the Solids Distribution in a Stirred Tank. IChemE Symp Series, No. 136, 1-8.

11.   Kilander, J. and Rasmuson, A. (2005) Energy Dissipation and Macro Instabilities in a Stirred Square Tank Investigated Using an LE PIV Approach and LDA Measurements. Chemical Engineering Science, 60, 6844-6856.http://dx.doi.org/10.1016/j.ces.2005.02.076

12.   Kresta, S.M., Mao, D. and Roussinova, V. (2006) Batch Blend Time in Square Stirred Tanks. Chemical Engineering Science, 61, 2823-2825.http://dx.doi.org/10.1016/j.ces.2005.10.069

13.   Kilander, J. Blomström, S. and Rasmuson, A. (2006) Rasmuson Spatial and Temporal Evolution of Floc Size Distribution in a Stirred Square Tank Investigated Using PIV and Image Analysis. Chemical Engineering Science, 61, 7651- 7667. http://dx.doi.org/10.1016/j.ces.2006.09.001

14.   Kilander, J. Blomström, S. and Rasmuson, A. (2007) Scale-Up Behaviour in Stirred Square Flocculation Tanks. Chemical Engineering Science, 62, 1606-1618.http://dx.doi.org/10.1016/j.ces.2006.06.002

15.   Brucato, A. Ciofalo, M., Grisafi, F. and Micale, G. (1998) Numerical Prediction of Flow Fields in Baffled Stirred Vessels: A Comparison of Alternative Modelling Approaches. Chemical Engineering Science, 53, 3653-3684. http://dx.doi.org/10.1016/S0009-2509(98)00149-3

16.   Aubin, J. Fletcher, D. and Xuereb, C. (2004) Modelling Turbulent Flow in Stirred Tanks with CFD: The Influence of the Modelling Approach, Turbulence Model and Numerical Schema. Experimental Thermal and Fluid Science, 28, 431-445.http://dx.doi.org/10.1016/j.expthermflusci.2003.04.001

17. Bakker, A. and Oshinowo, L.M. (2004) Modelling of Turbulence in Stirred Vessels Using Large Eddy Simulation. Chemical Engineering Research and Design, 82, 1169-1178.

18. Bakker, A. Oshinowo, L.M. and Marshall, E.M. (2000) The Use of Large Eddy Simulation to Eddy Simulation to Study Stirred Vessel Hydrodynamics. Proceeding of the 10th European Conference on Mixing, Delft, 2-5 July 2000, 247-254.

19. Patankar, S.V. (1980) Numerical Heat Transfer and Fluid Flow. McGraw Hill, New York.

Chapter 4

Discrete Tracer Point Method to Evaluate Turbulent Diffusion in Circular Pipe Flow

Arif Widiatmojo[1], Kyuro Sasaki[1], Nuhindro Priagung Widodo[2], and Yuichi Sugai[1]

[1]Department of Earth Resources Engineering, Faculty of Engineering,
Kyushu University, Fukuoka, Japan

[2]Department of Mining Engineering, Faculty of Mining and Petroleum
Engineering, Institut Teknologi Bandung, Bandung, Indonesia

ABSTRACT

Diffusion of a solute in turbulent flows through a circular pipe
or tunnel is an important aspect of environmental safety. In this
study, the diffusion coefficient of turbulent flow in circular pipe
has been simulated by the Discrete Tracer Point Method (DTPM).
The DTPM is a Lagrangian numerical method by a number of

imaginary point displacement which satisfy turbulent mixing by velocity fluctuations, Reynolds stress, average velocity profile and a turbulent stochastic model. Numerical simulation results of points' distribution by DTPM have been compared with the analytical solution for turbulent plug-flow. For the case of turbulent circular pipe flow, the appropriate DTPM calculation time step has been investigated using a constant β, which represents the ratio between average mixing lengths over diameter of circular pipe. The evaluated values of diffusion coefficient by DTPM have been found to be in good agreement with Taylor's analytical equation for turbulent circular pipe flow by giving β = 0.04 to 0.045. Further, history matching of experimental tracer gas measurement through turbulent smooth-straight pipe flow has been presented and the results showed good agreement.

INTRODUCTION

The diffusion of gas and other particulate matter in pipe or channel flows is important aspect to meet the safety requirements. It controls the longitudinal spreading and the residence time of gases or other particulate matter throughout the pipe. Diffusion occurred in turbulent flow in circular airway has been investigated for a century. Several researches were done by conducting experimental works or numerical approaches. When a pulsed substance or solute is injected into a straight pipe flow, it is advected and diffused to a relative reference moving with certain average velocity. Diffusion in the turbulent pipe flow is mainly characterized by axial velocity profile and velocity fluctuation in flow direction, because the radial gradient of solute concentration is much less than that of flow direction and also radial diffusion is limited by its pipe wall. Furthermore, turbulent mixing motions at different radial positions enhance the diffusion degree in flow direction.

Taylor [1, 2] and Aris [3] have made important contributions to develop theories about longitudinal diffusion in pipe flow. Taylor [2] also analyzed longitudinal diffusion in turbulent flow. He used the Reynolds analogy assuming radial diffusivity is analogous with heat and mass transfer in turbulent flow as well as transfer of fluid momentum. He neglected the contribution of molecular diffusion in both radial and axial directions which are negligibly small compared with turbulent eddy mixing diffusion in high Reynolds number. He also evaluated velocity at a certain distance from center of pipe and radial diffusivity as a function of universal velocity profile modified from Goldstein [4]. These assumptions are valid only for higher Reynolds number (Re > 2 × 10^4) where viscous sublayer and transition layer are negligibly thin. Taylor [2] also proposed the relationship between the longitudinal diffusion coefficient, E (m²/s), against pipe diameter and turbulent shear velocity given by following equations:

$$E = 5.05du*$$
(1)

$$u* = \sqrt{\frac{\tau}{\rho}} = U_m\sqrt{\frac{f}{8}}$$
(2)

Where, τ (Pa) and u* (m/s) are shear stress and friction velocity in an arbitrary sub layer of the flow, ρ (kg/m³) is fluid density, d (m) is pipe diameter, f (-) is DarcyWeisbach friction factor and U$_m$ (m/s) is cross sectional average velocity.

There are numbers of studies especially for atmospheric pollutant dispersion using random walk as basic method. Most of researches addressed the ideal homogenous turbulence. The pioneer was Taylor [5] who proposed continuous random walk theory for ideal homogeneous turbulent. For his random walk model, Pope [6] applied the well-known Langevin equation, a stochastic differential equation. He pointed out the consistency condition in the case of homogeneous turbulent satisfies linear Gaussian model while for inhomogeneous turbulent satisfies nonlinear Gaussian model. Milojevic [7] simulate particle dispersion in an ideal homogenous flow by incorporating both Lagrangian and Eulerian method. Further, he found high particle concentration at low fluctuation velocity and vice versa. Kroger and Drossinos [8] applied random walk method to simulate thermophoretic particles deposition in turbulent boundary layer using Lagrangian method. He considered velocity, temperature fields and thermophoretic force as Gaussian random fields of which the mean values were obtained from law-to-law wall reactions and from Knudsen number dependent expression of thermophoretic force. The root-mean-square (r.m.s) fluctuation was calculated by polynomial fit with experimental data. Luhar, Ashok and Britter [9] developed random walk for dispersion in a convective boundary layer in inhomogeneous flow. They incorporated well mixed conditions, skewness in vertical velocity and Gaussian random forcing.

Pulsed injection measurement method by using NaCl into water stream in smooth glass pipe was conducted by Sittel, Threadgill and Schnelle [10], while Taylor [2] conducted both in smooth and artificially roughened glass pipes. Furthermore, they applied least square fitting method for their measurements, and showed higher prediction values compared to Taylor's Equation (1). Higher values of diffusion coefficient were also observed by Hull and Kent [11] by using radioactive tracer injected into a long oil pipelines. Their reason was because of pipe bends and variation in elevation due to terrain profile.

Measurements of diffusion of gas-phase in smooth pipe flows were carried out by Keyes [12], while Davidson, Farqurharson, Picken and Taylor [13] conducted in a rough and long pipeline. Widodo, Sasaki, Gautama and Risono [14] also conducted tracer gas measurement in an underground mine ventilation airways. They found higher diffusion coefficient for the airways flow than those evaluated using Equation (1) for smooth pipe. This phenomenon is well explained by several researchers [15-17]. Compared with obtained data using a solute in water flow, data provided by Keyes [12] and Taylor [2] were measured at relatively lower Reynolds number region. They concluded that the effect of molecular diffusion cannot be neglected in low Reynolds number. Furthermore, because liquid has less molecular transfer, it has a higher Schmidt number compared to gas. Tichacek et al. [17] improved Taylor's model by considering the molecular diffusion and mean velocity profile based on averaging velocity profiles. They reported that their calculation result deviates significantly for Reynolds number about 4.2 ´ 10⁴, due to the different sets of velocity profile. Thus, the velocity profile used in calculations has a sensitive effect on evaluated values of diffusion coefficient. Figure 1 shows evaluated diffusion coefficients reported in several researches. The empirical relationship proposed by Wen and Fan [18] is also plotted in Figure 1 and shows higher prediction compared to [10] and [17]. Taylor's analytical solution (Equation (1)) is also plotted together with 10% error bars. All have shown the scattered result to each other. Further, it is still necessary to develop further numerical method for gas diffusion in mine airways to simulate the longer travelling time and high diffusion coefficient than those for circular tube flow as demonstrated by Sasaki, Widiatmojo, Arpa and Sugai [19].

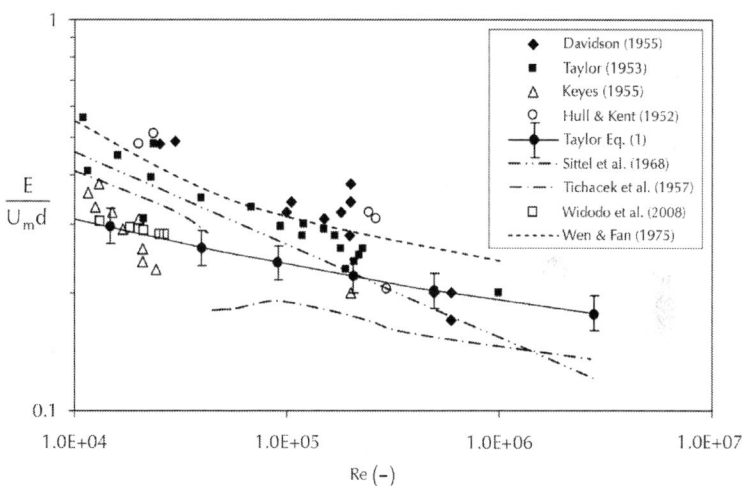

Figure 1: Measurement data, empirical and analytical relationship of longitudinal diffusion coefficient versus Reynolds number from various studies.

The advantages of proposed DTPM are that the calculations of concentration gradient in space or time domain, which are commonly employed in numerical simulations, are not required. It may reduce the computational time. It is also free of grid requirement and the visualization of points' distribution is simple.

NUMERICAL MODELS FORMULATIONS

Point Movement

The scheme of present numerical simulations has been developed by moving points with regards to velocity profile and turbulent intensities in a turbulent circular pipe flow, depends on radial position of the point in circular cross section. Figure 2 shows schematic variable definitions on the numerical calculation scheme using Cartesian coordinate (x, y, z) to describe tracer positions in circular airway with radius R or diameter d (=2R). The x is distance from the initial position in flow (longitudinal) direction, r is radial position from tube center perpendicular to wall surface, and ϕ is tangential direction perpendicular to x and r. As for fully developed turbulent flow, mean velocities in r and ϕ directions can be zero.

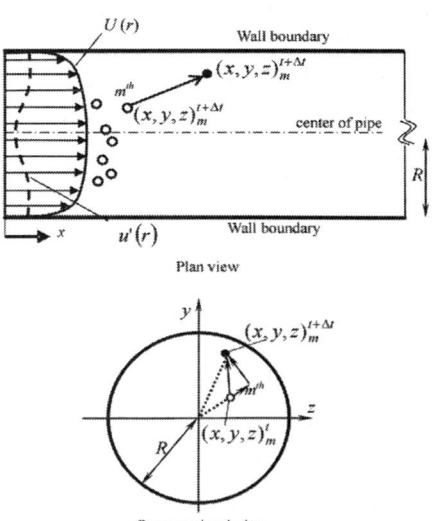

Figure 2: Schematic definition of tracer point movements in pipe flow (x-y and y-z cross sections).

Turbulent flow is characterized by its stochastic properties. The velocity at a given specific position fluctuates around its mean value. The velocity fluctuation intensity is known as root mean square (r.m.s) values which vary as function of r. The time of averaged flow velocities and the turbulent intensities at a certain position in cylindrical coordinate system (x, r, ϕ) are defined as

(U, 0, 0) and (u', v_r', v_ϕ') respectively. The moving of points may be treated easily on Cartesian coordinate system (x, y, z), therefore, in present calculations, turbulent intensities in (x, r, ϕ) directions are transformed into (x, y, z) directions of Cartesian coordinate.

Assumed instantaneous turbulent fluctuations are (u', v', w') in (x, y, z) directions, these can be obtained by transforming velocity fluctuations (u', v_r', v_ϕ') as follows (see Figure 3);

$$u' = u'$$
$$v' = v_r' \sin \phi + v_\phi' \cos \phi$$
$$w' = v_r' \cos \phi - v_\phi' \sin \phi$$

(3)

Suppose the position of mth point dosed into the flow at origin is denoted with superscript showing the elapsed time, t = 0, its moving distances ($\Delta x, \Delta y, \Delta z$) during time step Δt are given by;

$$\left(\Delta x, \Delta y, \Delta z \right)_m^t = \left(u, v, w \right)_m^t \Delta t = \left(U + u', v', w' \right)_m^t \Delta t$$

(4)

Its displacement is expressed at t + Δt and t by;

$$\left(x, y, z \right)_m^{t+\Delta t} = \left(x, y, z \right)_m^t + \left(\Delta x, \Delta y, \Delta z \right)_m^t$$

(5)

Laufer [20] conducted measurements of turbulent intensities and Reynolds stress of air flowing in a straightsmooth pipe with diameter, d = 0.254 m. Measured r.m.s values of turbulent intensities expressed by ($v_{r(rms)}', v_{\phi(rms)}', u_{(rms)}'$) in each axis (r, ϕ, x) were presented as ratios against u* as function of r/R. In present numerical simulations, the polynomial approximations for Laufer's measurements (see Figure 4) were used to calculate the turbulent intensities at a point's position, r/R.

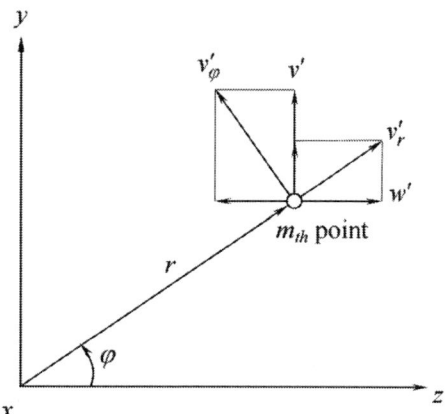

Figure 3: Coordinate transformation of velocity vectors in r and φ into y and z direction (Cartesian coordinate).

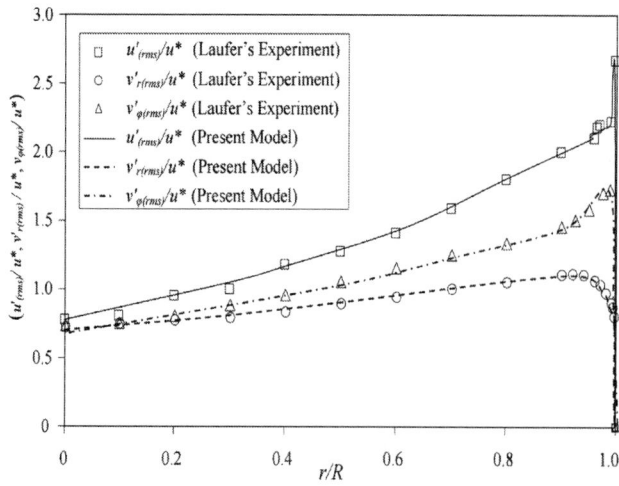

Figure 4: Turbulent model of r.m.s value of turbulent intensities (u', v'_r, v'_φ) compared with Laufer's results.

Average Velocity Profile

If dimensionless value of longitudinal average velocity is defined as:

$$U^+ = \frac{U}{u^*}$$

(6)

Dimensionless distance from the wall is given by:

$$y^+ = \frac{(R-r)}{\upsilon} u^*$$

(7)

Where υ (m²/s) is kinematic viscosity. According to Kenyon [21], the relationships of U^+ and y^+ have been presented by Nikuradse with equations for three regions, that are viscous sub layer ($y^+ \leq 5$), buffer layer ($5 < y^+ \leq 30$) and turbulent zone ($y^+ > 30$).

$$U^+ = y^+, \quad \text{for} \quad y^+ \leq 5$$

(8)

$$U^+ = 5.0 \ln y^+ - 3.05, \quad \text{for} \quad 5 < y^+ \leq 30$$

(9)

$$U^+ = 2.5 \ln y^+ + 5.5, \quad \text{for} \quad y^+ > 30$$

(10)

Equations (8) to (10) have been confirmed to agree well with equations proposed by Reichardt [22]. In present simulations, these equations were used to calculate axial velocity, U(r) of each point.

Turbulent Stochastic Model

A stochastic approach was applied to determine points' diffusion process in the turbulent flow. In present numerical model, it is supposed that the point movements were based on turbulent eddy motion, which satisfies Gaussian probability density function (hereinafter GPDF) with a standard deviation equal to the turbulent intensities or the r.m.s value of velocity fluctuations in each direction (see Rouse [23]). In the simulations, three pseudo-random numbers follow GPDF were generated using Box Muller algorithm to calculate turbulent velocity vector (v'_r, v'_φ, u').

As described previously, Laufer's measurement results of turbulent intensities were presented in normalized values over shear velocity, u^*, which is function of cross sectional average velocity, U_m, and friction factor, f. The relationship between f and Re was presented with empirical equation by Colebrook [24];

$$\frac{1}{\sqrt{f}} = 2.0\log\left(\mathrm{Re}\sqrt{f}\right) - 0.8$$

(11)

Turbulent Reynolds Stress Model

In turbulent shear flow, fluid particles are translated from slower region to faster one and Reynolds stress is generated. It shows a time-averaged correlation between longitudinal and radial velocity fluctuations in the shear flows. In present study, effects of Reynolds stress correlations have been investigated by giving relationship between x axis velocity fluctuation, u', with velocity fluctuation in r direction, v'_r, expressed as;

$$-u'v'_r > 0$$

(12)

In the simulations, the value of u' including its sign was firstly given as a random number following GPDF described previously in preceding chapter, then absolute values of $|v'_r|$ were generated using similar method, but the signs of u' were decided to satisfy Equation (12).

Boundary Condition and Initial Condition

In fact, point's displacement near wall and its wall interaction has not been well understood to simulate breaking sub-layer. The boundary condition for Lagrangian random walk still needs a calculation model. Several boundary models have been proposed in previous studies. Those methods depend on the physical and numerical factors considered in the simulation [25].

In present DTPM simulations, the reflection boundary condition at the airway wall is satisfied by a repositioning numerical treatment if points are out of the flow region. Thus, by this boundary condition the zero-flux condition at wall can be satisfied. A reflection boundary scheme is modeled as shown in Figure 5 and is used in present simulation. Similar boundary model was also proposed by Szymczak and Ladd [26, 27].

Another boundary condition namely rearrangement model has been investigated. In this model, random number is continuously generated until the point is repositioned within the flow regime. Szymczak and Ladd [26] also reviewed this kind of boundary condition as multiple rejection method. Figure 6 shows the comparison of two models. It can be seen that the reflection model resulted in higher dispersion than the rearrangement model. The reason is because the reflection model allows higher possibilities for points to be repositioned at near wall region. These movements make higher "retaining effect" in low velocity layer near wall. Further simulations indicated that the reflection model shows closer results to Taylor's analytical solution. Thus, reflection model is adapted for present DTPM numerical simulations.

Other zero-flux boundary conditions also have been proposed. Drazer and Koplik [28]; Kurowski, Ippolito, Hulin, Koplik and Hinch [29] proposed rejection boundary where the points do not change its position in the given time step and other random diffusivity is recalculated until the movement satisfy $r < R$. Salles et al. [30] and Maier et al. [31] suggested interruption boundary condition at which points stop at wall and its time is incremented by modified time step, λDt where $\lambda = 0 \sim 1$ is adjustment factor. However, only the reflection method imposes the zero-flux boundary condition while other boundary conditions lead to incorrect concentration profiles near wall boundary [26].

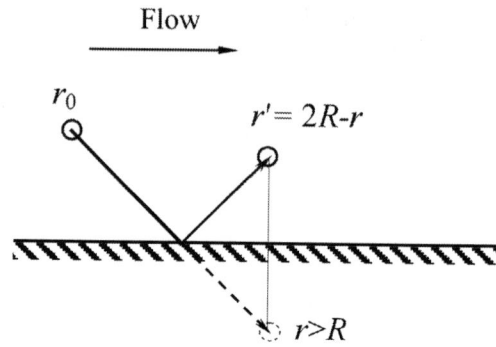

Figure 5: Numerical boundary condition at wall, a reflection model.

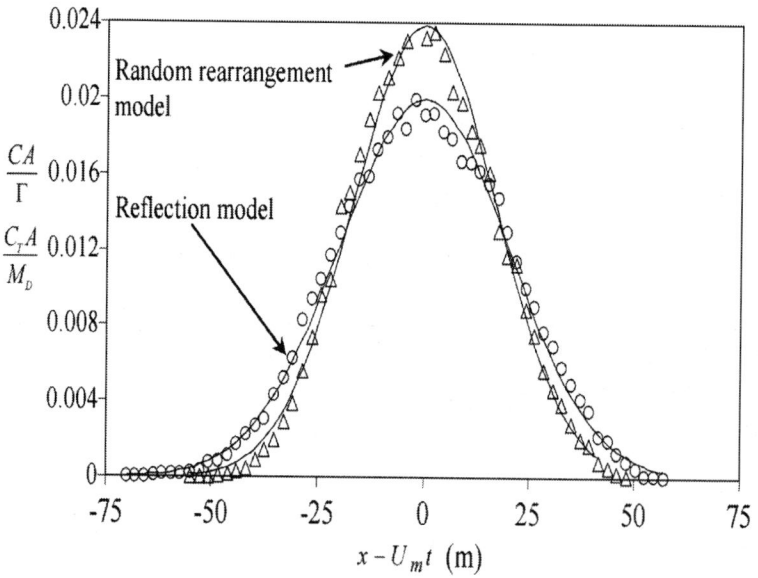

Figure 6: Comparison of numerical result of DTPM after t = 100 s by us-ing different wall boundary conditions (Dt = 0.5 s, U_m = 4 m/s, d = 2 m).

CALCULATION RESULTS AND EVALUATION OF DIFFUSION COEFFICIENT

One-Dimensional Diffusion

Taylor [1, 2] proposed that concentration distribution of solute after certain time, t, is symmetrically distributed follows the partial differential equation. He proposed concentration gradient at x and r direction which move to x direction with constant average cross sectional velocity, U_m;

$$\frac{\partial C}{\partial t} + U_m \frac{\partial C}{\partial x} = E\left(\frac{\partial^2 C}{\partial r^2} + \frac{1}{r}\frac{\partial C}{\partial r} + \frac{\partial^2 C}{\partial x^2}\right)$$

(13)

Several studies (see Wen and Fan [18]) also proposed similar numerical expression of solute diffusion plug flow model.

The solution of Equation (13) can be obtained by assuming that molecular diffusion is neglected and concentration gradient in radial direction is negligible. The variable, E shows effective diffusion coefficient in axial direction. Since the center of dispersed solute is assumed to be at $x = U_m t$ after elapsed time, t, and the solution of Equation (13) can be given with an equation similar with Gaussian distribution:

$$\frac{A \cdot C(x,t)}{\Gamma} = \frac{1}{2\sqrt{\pi E t}} \exp\left(\frac{-(x - U_m t)^2}{4Et}\right)$$

(14)

Where G (m³) is total volume of released solute at x = 0, and A (m²) is cross sectional area of pipe.

In the DTPM simulation, concentration of diffused points, C, can be calculated by;

$$C_T = \frac{\Delta m}{\Delta V} = \frac{\Delta m}{\Delta x \pi R^2}$$

(15)

Where Δm is number of point counted in the numerical control volume which located at a certain downstream position, given by $\Delta V = \Delta X \pi d^2/4$ and $\Delta X = U_m \Delta t$. Suppose the total number of points released from the origin is M_D, The normalized concentration of points is expressed as $C_T A/M_D$.

Simulation of Turbulent Flow through Circular Airway

The DTPM simulations on turbulent flow of a straightsmooth airway have been carried out. Figure 7 shows the evolution of point's distribution after several elapsed time since released into the flow (U_m = 1.5 m/s, d = 0.5 m). It can be observed that the asymmetry of point's distribution is gradually diminished as the travelled distance increases.

Figures 8 and 9 show the results of simulation for flow with d = 2 m, U_m = 4 m/s (f = 0.01315) and d = 4 m, U_m = 5 m/s (f = 0.0112) respectively after t = 100 s. It can be inferred that different calculation time step, Δt, gives different evaluated value of effective diffusion coefficient. To consider this effect, the evaluated value of effective diffusion coefficient is plotted against dimensionless value, β (-) defined as:

$$\beta = \frac{(u'_{rms})_{r=0} \Delta t}{d}$$

(16)

Where $(u'_{rms})_{r=0}$ is r.m.s value of streamwise velocity fluctuation in the center of pipe. It may be possible to regard β as ratio of the average mixing length to pipe diameter.

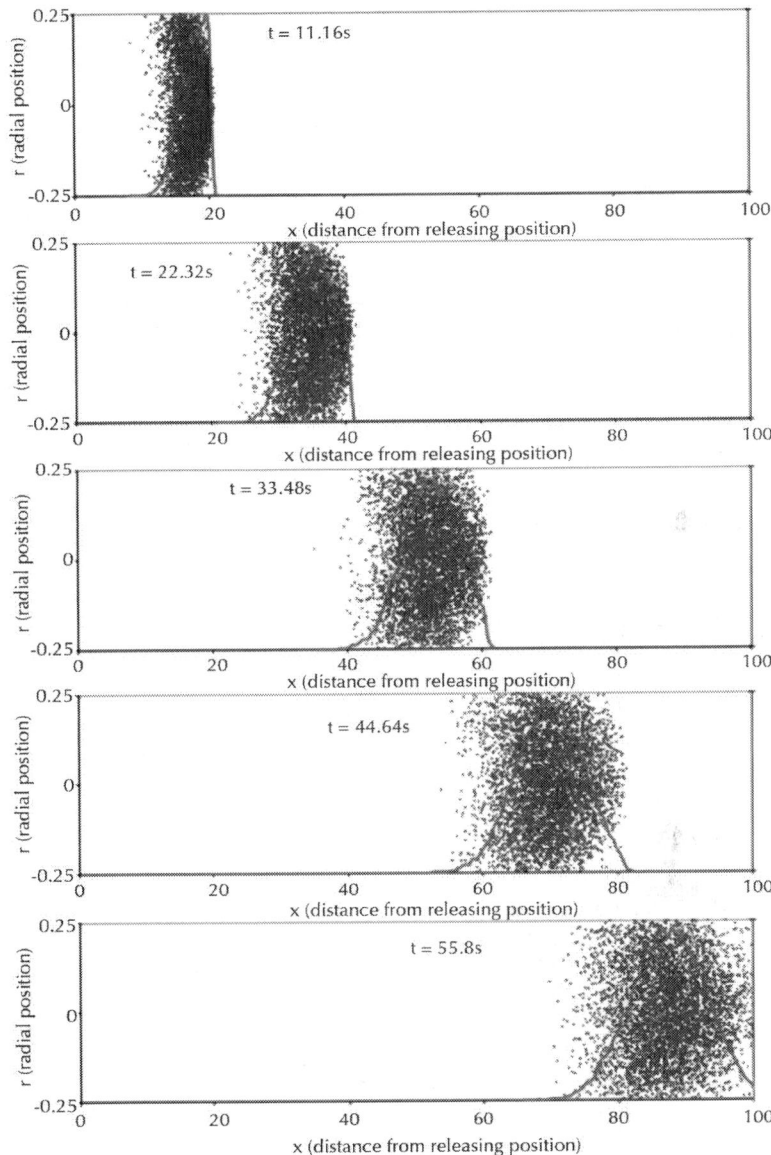

Figure 7: Points distribution showing diffusion in longitudinal direction for U_m = 1.5 m/s, d = 0.5 m, Δt = 1 s, f = 0.0211, after five consecutive elapsed time (horizontal axis and vertical axis is travelled distance in flow direction and radial distribution respectively).

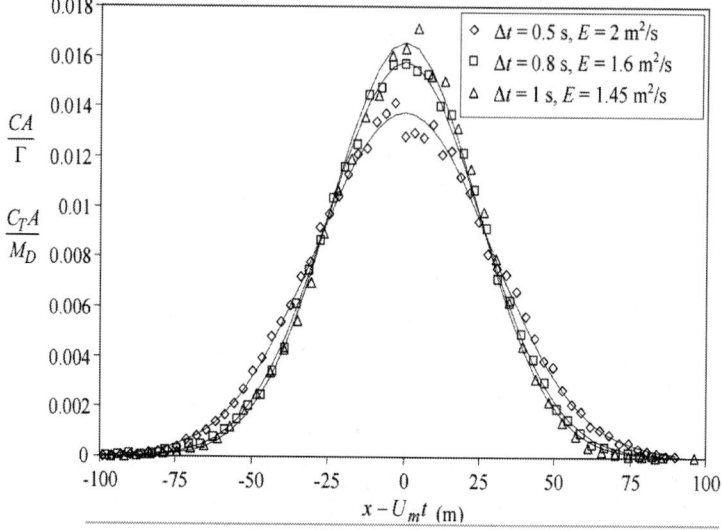

Figure 8: Results of DTPM simulation with different Δt at t = 200 s for circular channel flow with U_m = 4 m/s, d = 2 m (f = 0.01315).

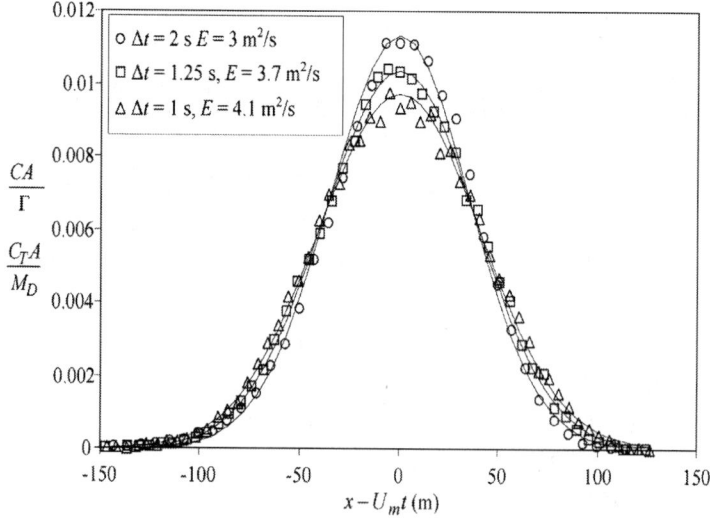

Figure 9: Results of DTPM simulation with different Δt at t = 200 s for circular channel flow with U_m = 5 m/s, d = 4 m (f = 0.0112).

Figures 10(a)-(d) shows evaluated values of effective diffusion coefficient of different time step for Re = 5 × 10^5, 7.5 × 10^5, 9.4 × 10^5 and 1.25 × 10^6. The values of E calculated by Equation (1) are also presented as comparisons. From the results, it can be seen that the evaluated values of E from present DTPM for β = 0.04 ~ 0.045 intersect with those Taylor analytical solution given by Equation (1).

The results of evaluated E of DTPM showing relatively non-linear and inversely proportional correlation to the value of β, before gradually attain constant value as higher value of Δt is applied. Further simulations were done by setting β = 0.040, 0.043 and 0.045 at different flow conditions. As shown in Figure 11, the results indicate linear relationship between Taylor's analytical equation and ones evaluated by DTPM. The value of β = 0.043 used for DTPM simulations is appropriate in order to get longitudinal diffusion coefficient, E, which has good agreement with Taylor's analytical solution.

SIMULATION OF TRACER GAS EXPERIMENT THROUGH SMOOTH-CIRCULAR PIP

Tracer gas diffusion experiment was conducted at a laboratory scale by Widodo [32]. The experimental apparatus mainly consists of smooth circular pipe as scaled pipe, gas injection apparatus, and gas measurement apparatus.

For the straight airway, he used a pipe with smooth lining 0.025 m diameter, 30 m length and placed horizontally. Tracer gas was released by breaking balloons filled with methane (CH_4) and measured arriving concentration at the end of pipe. Methane concentration was measured by an original infrared adsorption gas detector using 3.4 mm He-Ne laser with an infrared (IR) sensor, air pump, air mass flow meter and amplifier as shown in Figure 12. Sampling rate was set as the maximum flow rate at which the gas sensor could still detect the gas concentration correctly. The data were recorded with a data logger connected to a PC.

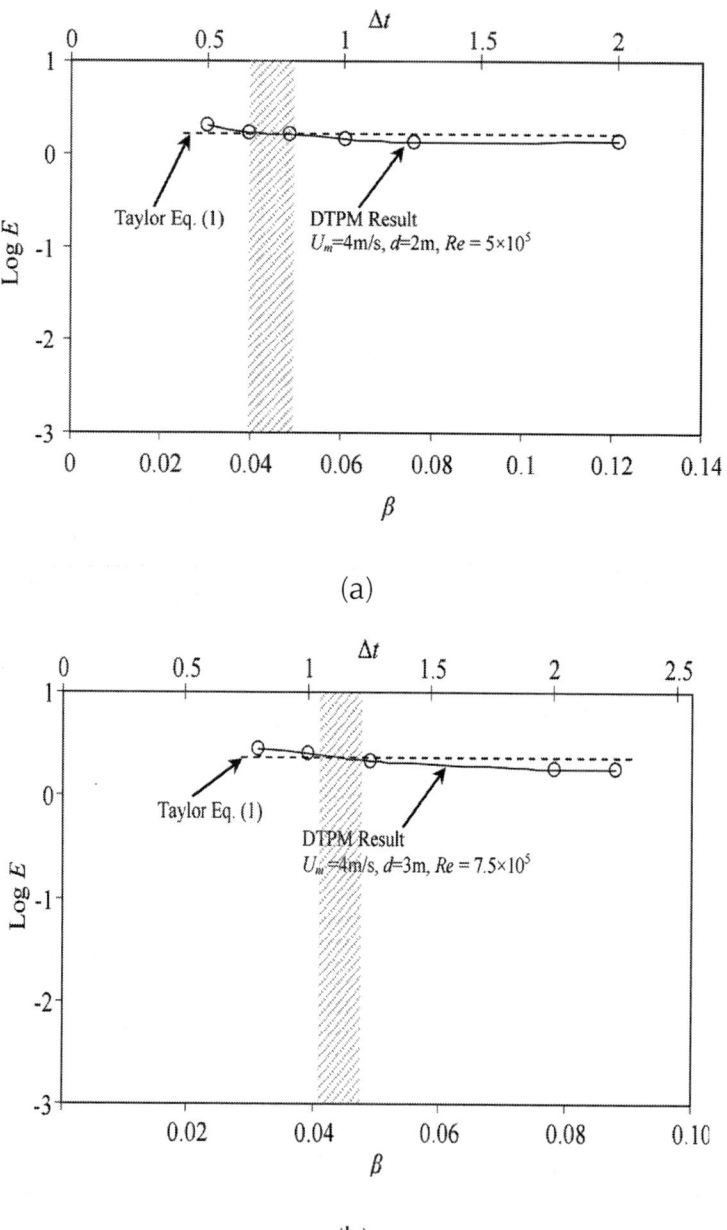

(a)

(b)

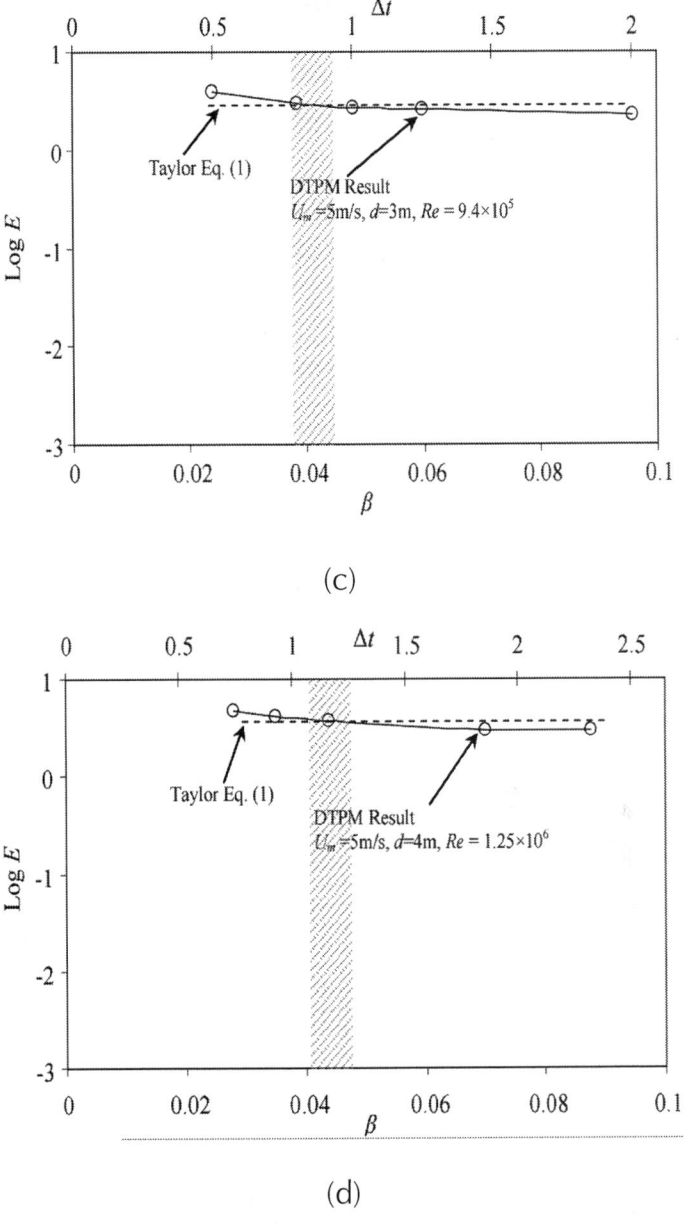

(c)

(d)

Figure 10: Effect of dimensionless value, β, on evaluated diffusion coefficient, E, for different Reynolds number compared with Taylor's analytical solution given by Equation (1). (a) Re = 5 × 10⁵; (b) Re = 7.5 × 10⁵; (c) Re = 9.4 × 10⁵; (d) Re = 1.25 × 10⁶.

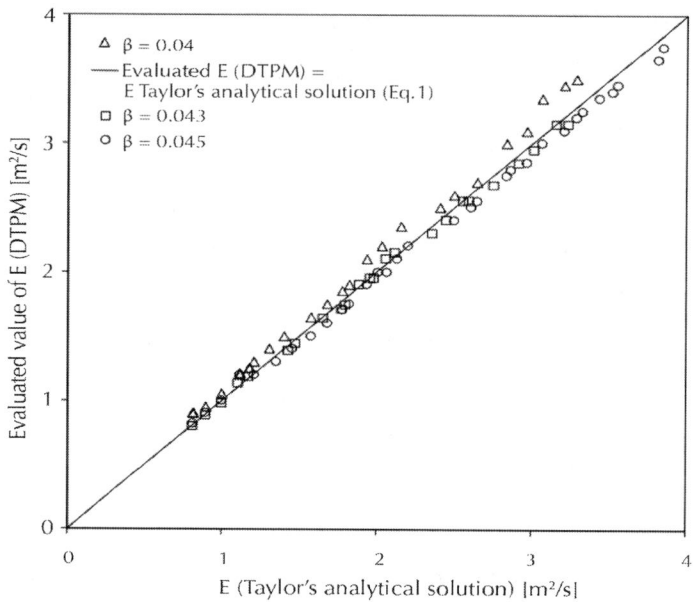

Figure 11: Evaluated result of diffusion coefficient evaluated by DTPM considering β = 0.04, 0.043 and 0.045 compared with Taylor solution of (Equation (1)).

Beforehand, the conversion of voltage reading by data logger to gas concentration was calibrated based on calibration data with the standard methane-air mixture. The validity of concentration reading from IR sensor was crosschecked using gas chromatograph and showed good linear fit.

The measurement was conducted for Reynolds number, Re = 6085 (U_m = 3.87 m/s, d = 2R = 0.025 m). Measurements for higher Re were also done; however, the data were inadequately acquired due to insufficient sampling rate to compensate higher flow velocity.

For the DTPM simulation, calculation parameters were decided based on the experimental properties such as pipe's length, average cross sectional velocity and pipe's diameter. Value of friction factor, f, was calculated using Equation (11) and calculation time step was defined using Equation (16) as Dt = 0.0092 s. To verify the effect of initial points' distribution, three different conditions were considered; 1) point source ($x_0 = 0$, $r_0 = 0$); 2) uniformly distributed at $x_0 = 0$ ($0 \leq r_0 < R$); 3) uniformly distributed at $x_0 \pm 2d$.

Figure 13 shows the evolution of radial and longitudinal point's distribution in five consecutive times for point source initial condition. The results shown in Figure 14 implied that the difference of initial points' distribution has no significant effect on result of DTPM. The length of flow domain is long enough for the radial diffusivity to attain radial homogeneity of flow. It may also imply that the sudden break of balloons during release has less effect on the arriving concentration as the measurement distance is long compared to diameter d>> pipe length. In general, the results confirm that present DTPM is able to simulate the turbulent in straight and smooth circular pipe.

In the real case, the utilization of tracer gas measurement is not merely straight pipe, but also tunnels network [19, 32-35]. This kind of network is assembled in the form of interconnecting tunnels which allow airflow to be separated or rejoined at junction. The mechanism of points tracking as proposed in this study can be developed to consider network flow by combining with a scheme to treat flow separation. The developed scheme is supposed to be able to simulate point's distribution at arbitrary position in the network and allow easy dispersion evaluation of gas or other particulates spreading.

CONCLUSIONS

In this study, effective diffusion coefficients of turbulent flow in circular pipe or channel have been evaluated by the Discrete Tracer Point Method (DTPM), a Lagrangian numerical simulation method. The present study is summarized as follows:

- DTPM simulation procedures have been presented to simulate the displacement of points released into straightsmooth circular airway flow by giving turbulent average velocity, intensity of velocity fluctuation and Reynolds stress;

- A simple procedure to represent diffusion of points in the airway flow has been employed by generating random number which satisfies Gaussian probability functions with value of turbulent intensities in each direction as standard deviations;

- Appropriate calculation time step was proposed by matching with Taylor's analytical equation by considering the ratio between the average mixing lengths to airway's diameter, **β** @ 0.043;

- The result of matching curve of DTPM with experimental result by considering different initial points distribution has shown that the present DTPM can be used to simulate turbulent advection-diffusion in smoothstraight pipe representing the ideal model of straight pipe or channel;

- Present DTPM simulation has promising possibility to simulate the network flow by combining with a scheme to treat flow separation at junction.

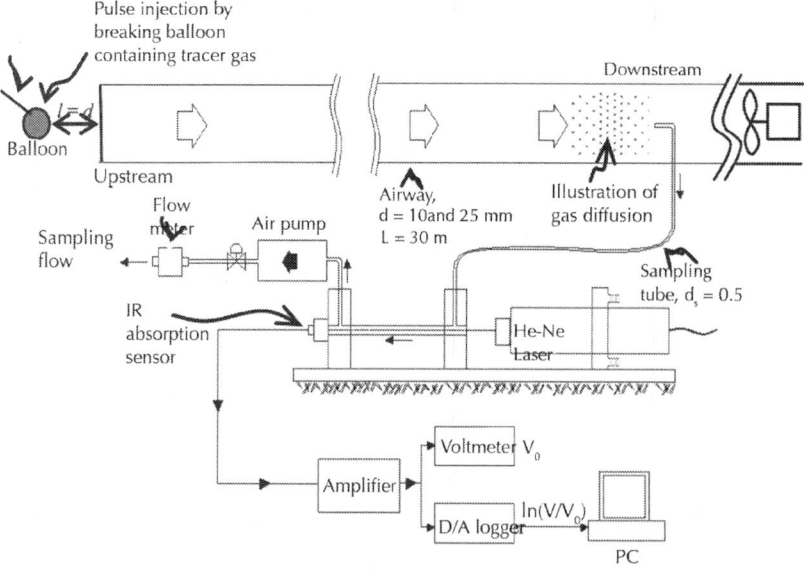

Figure 12: Experimental setting for turbulent diffusion through circular smooth-straight pipe (after Widodo [32]).

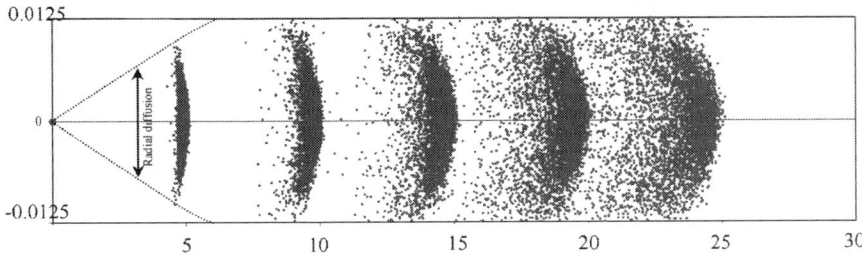

Figure 13: Point distribution after t = 1.08 s, t = 2.16 s, t = 3.24 s, t = 4.32 s, t = 5.40 s with point source initial condition; (U_m = 3.87 m/s, d = 0.025 m, Δt = 0.0092 s, f = 0.0211, horizontal axis and vertical axis is travelled distance in flow direction and radial distribution respectively).

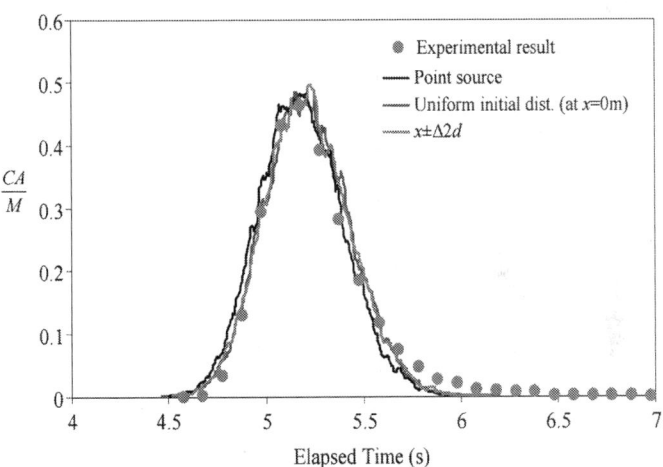

Figure 14: Simulation results with different initial condition of point's distribution for U_m = 3.87, d = 0.025 m, f = 0.035.

ACKNOWLEDGEMENTS

Authors would like to thank Japan Ministry of Education, Culture, Sports, Science and Technology and Kyushu University Global COE program for the financial support.

REFERENCES

1.   G. I. Taylor, "Dispersion of Soluble Matter in Solvent Flowing Slowly through a Tube," Proceedings of the Royal Society of London. Series A. Mathematical and Physical Sciences, Vol. 219, No. 1137, 1953, pp. 186-203. doi:10.1098/rspa.1953.0139

2.   G. I. Taylor, "The Dispersion of Matter in Turbulent Flow through a Pipe," Proceedings of the Royal Society of London. Series A. Mathematical and Physical Sciences, Vol. 223, No. 1155, 1954, pp. 446-468. doi:10.1098/rspa.1954.0130

3.   R. Aris, "On the Dispersion of a Solute in a Fluid Flowing through a Tube," Proceedings of the Royal Society of London. Series A. Mathematical and Physical Sciences, Vol. 235, No. 1200, 1956, pp. 67-77. doi:10.1098/rspa.1956.0065

4.   S. Goldstein, "Modern Developments in Fluid Dynamics," Dover, New York, 1965.

5.   G. I. Taylor, "Diffusion by Continuous Movements," Proceedings of London Mathematical Society, Vol. 20, No. 1, 1954, pp. 196-211. doi:10.1112/plms/s2-20.1.196

6.   S. B. Pope, "Consistency Conditions for Random-Walk Models of Turbulent Dispersion," Physics of Fluids, Vol. 30, No. 8, 1987, pp. 2374-2379. doi:10.1063/1.866127

7.   D. Milojevic, "Lagrangian Stochastic-Deterministic Predictions of Particle Dispersion in Turbulence," Particle & Particle Systems Characterization, Vol. 7, No. 1-4, 1990, pp. 181-190. doi:10.1002/ppsc.19900070132

8.   C. Kröger and Y. Drossinos, "A Random-Walk Simulation of Thermophoretic Particle Deposition in a Turbulent Boundary Layer," International Journal of Multiphase Flow, Vol. 26, No. 8, 2000, pp. 1325-1350. doi:10.1016/S0301-9322(99)00092-0

9.   A. K. Luhar and R. E. Britter, "A Random Walk Model for Dispersion in Inhomogeneous Turbulence in a Convective

Boundary Layer," Atmospheric Environment, Vol. 23, No. 9, 1989, pp. 1911-1924. doi:10.1016/0004-6981(89)90516-7

10. C. N. Sittel Jr., W. D. Threadgill and K. B. Schnelle Jr., "Longitudinal Dispersion for Turbulent Flow in Pipes," Industrial & Engineering Chemistry Fundamentals, Vol. 7, No. 1, 1968, pp. 39-43. doi:10.1021/i160025a007

11. D. E. Hull and J. W. Kent, "Radioactive Tracers to Mark Interfaces and Measure Intermixing in Pipelines," Industrial & Engineering Chemistry, Vol. 44, No. 11, 1952, pp. 2745-2750. doi:10.1021/ie50515a066

12. J. J. Keyes Jr., "Diffusional Film Characteristics in Turbulent Flow: Dynamic Response Metho," AiChE Journal, Vol. 1, No. 3, 1955, pp. 305-311. doi:10.1002/aic.690010306

13. F. Davidson, D. C. Farqurharson, J. Q. Picken and D. C. Taylor, "Gas Mixing in Long Pipelines," Chemical Engineering Science, Vol. 4, No. 5, 1955, pp. 201-205.doi:10.1016/0009-2509(55)80006-1

14. N. P. Widodo, K. R. Sasaki, R. S. Gautama and Risono, "Mine Ventilation Measurements with Tracer Gas Method and Evaluations of Turbulent Diffusion Coefficient," International Journal of Mining, Reclamation and Environment, Vol. 22, No. 1, 2008, pp. 60-69.doi:10.1080/17480930701474869

15. W. B. Krantz and D. T. Wasan, "Axial Dispersion in the Turbulent Flow of Power-Law Fluids in Straight Tubes," Industrial & Engineering Chemistry Fundamentals, Vol. 13, No. 1, 1974, pp. 56-62. doi:10.1021/i160049a011

16. O. Levenspiel, "Longitudinal Mixing of Fluids Flowing in Circular Pipes," Industrial & Engineering Chemistry, Vol. 50, No. 3, 1958, pp. 343-346. doi:10.1021/ie50579a034

17. L. J. Tichacek, C. H. Barkelew and T. Baron "Axial Mixing in Pipes," AIChE Journal, Vol. 3, No. 4, 1957, pp. 439- 442. doi:10.1002/aic.690030404

18. C. Y. Wen and L. T. Fan, "Models for Flow System and Chemical Reactors," Dekker, New York, 1975.

19. K. Sasaki, A. Widiatmojo, G. Arpa and Y. Sugai, "Airflow Measurements and Evaluation of Effective Diffusion Coefficient in Large Scale of Mine Ventilation Network Using with Tracer Gas Method," Journal of the Mining and Materials Processing Institute of Japan, Vol. 125, No. 12, 2009, pp. 614-620. doi:10.2473/journalofmmij.125.614

20. J. Laufer, "The Structure of Turbulent in Fully Developed Pipe Flow," Technical Report 1174, National Committee for Aeronautics, 1954. http://naca.central.cranfield.ac.uk/report.php?NID=5843

21. R. A. Kenyon, "Principles of Fluid Mechanics," Ronald Press, New York, 1960.

22. H. Reichardt, "Complete Representation of the Turbulent Velocity Distribution in Smooth Pipes," Journal of Applied Mathematics and Mechanics, Vol. 31. No. 7, 1951, pp. 208-219. doi:10.1002/zamm.19510310704

23. H. Rouse, "Advanced Mechanics of Fluids," John Wiley and Sons, Inc., New York, 1959.

24. C. F. Colebrook, "Turbulent Flow in Pipes, with Particular Reference to the Transition Region between the Smooth and Rough Pipe Laws," Journal of the ICE, Vol. 11, No. 4, 1939, pp. 133-156. doi:10.1680/ijoti.1939.13150

25. D. J. Thomson, W. L. Physick and R. H. Maryon, "Treatment of Interfaces in Random Walk Dispersion Models," Journal of Applied Meteorology, Vol. 36, No. 9, 1997, pp. 1284-1285. doi:10.1175/1520-0450(1997)036<1284:TOIIRW>2.0.CO;2

26. P. Szymczak and A. J. C. Ladd, "Boundary Conditions for Stochastic Solutions of the Convection-Diffusion Equation," Physical Review E, Vol. 68, No. 3, 2003, Article ID: 036704. doi:10.1103/PhysRevE.68.036704

27. P. Szymczak and A. J. C. Ladd, "Stochastic Boundary Conditions to the Convection-Diffusion Equation including Chemical Reactions at Solid Surfaces," Physical Review E, Vol. 69, No. 3, 2004, Article ID: 036704. doi:10.1103/PhysRevE.69.036704

28. G. Drazer and J. Koplik, "Tracer Dispersion in Two-Dimensional Rough Fractures," Physical Review E, Vol. 63, No. 5, 2001, Article ID: 056104.doi:10.1103/PhysRevE.63.056104

29. P. Kurowski, I. Ippolito, J. P. Hulin, J. Koplik and E. J. Hinch, "Anomalous Dispersion in a Dipole Flow Geometry," Physics of Fluids, Vol. 6, No. 1, 1994, pp. 108-117. doi:10.1063/1.868075

30. J. Salles, J. F. Thovert, R. Delannay, L. Prevors, J.-L. Auriault and P. M. Adler, "Taylor Dispersion in Porous Media. Determination of the Dispersion Tensor," Physics of Fluids A: Fluid Dynamics, Vol. 5, No. 10, 1993, pp. 2348-2376. doi:10.1063/1.858751

31. R. S. Maier, D. M. Kroll, R. S. Bernard, S. E. Howington, J. F. Peters and H. T Davis, "Pore-Scale Simulation of Dispersion," Physics of Fluids, Vol. 12, No. 8, 2000, pp. 2065-2079. doi:10.1063/1.870452

32. N. P. Widodo, "Study on Tracer Gas Method for Mine Ventilation Measurement and Evaluation of Gas Diffusion Coefficient," Ph.D. Thesis, Kyushu University, Fukuoka, 2008.

33. G. Xu, K. D. Luxbacher, S. Ragab and S. Schafrik, "Development of a Remote Analysis Method for Underground Ventilation Systems Using Tracer Gas and CFD in a Simplified Laboratory Apparatus," Tunneling and Underground Space Technology, Vol. 33, 2013, pp. 1-11. doi:10.1016/j.tust.2012.09.001

34. M. H. Johnson, Z. Zhai and M. Krarti, "Performance Evaluation of Network Airflow Models for Natural Ventilation," HVAC&R Research, Vol. 18, No. 3, 2012, pp. 349-365.

35. R. Gao, A. Li, X. Hao, W. Lei and B. Deng, "Prediction of the Spread of Smoke in a Huge Transit Terminal Subway Station under Six Different Fire Scenarios," Tunnelling and Underground Space Technology, Vol. 21, 2012, pp. 128-138. doi:10.1016/j.tust.2012.04.013

Electrical Capacitance Probe Characterization in Vertical Annular Two-Phase Flow

Grazia Monni[1], Mario De Salve[1], Bruno Panella[1], and Carlo Randaccio[2]

[1]Dipartimento Energia, Politecnico di Torino, Corso Duca degli Abruzzi 24, 10129 Torino, Italy

[2]SIET S.p.A., Via Nino Bixio 27, 29121 Piacenza, Italy

ABSTRACT

The paper presents the experimental analysis and the characterization of an electrical capacitance probe (ECP) that has been developed at the SIET Italian Company, for the measurement of two-phase flow parameters during the experimental simulation of nuclear accidents, as LOCA. The ECP is used to investigate a vertical air/

water flow, characterized by void fraction higher than 95%, with mass flow rates ranging from 0.094 to 0.15 kg/s for air and from 0.002 to 0.021 kg/s for water, corresponding to an annular flow pattern. From the ECP signals, the electrode shape functions (i.e., the signals as a function of electrode distances) in single- and two-phase flows are obtained. The dependence of the signal on the void fraction is derived and the liquid film thickness and the phase's velocity are evaluated by means of rather simple models. The experimental analysis allows one to characterize the ECP, showing the advantages and the drawbacks of this technique for the two-phase flow characterization at high void fraction.

INTRODUCTION

The design of new nuclear reactors requires carrying out integral and separate effect tests on simulation facilities, as well as performing safety systems verification and safety code validation. Within the framework of an Italian R&D program on nuclear fission, managed by ENEA and supported by the Ministry of Economic Development, the SPES3 (pressurized simulator for safety experiments) experimental facility, described by Carelli et al. [1], able to simulate the innovative small and medium size (PWR pressurized water reactor) nuclear reactors, is being built and will be operated at the SIET Company laboratories in Piacenza, Italy. In such facility, some design and beyond design basis accidents, like loss of coolant accidents (LOCAs), with and without the emergency heat removal systems, will be simulated [2]. The experimental data, concerning a series of primary and secondary loops breaks, will be fundamental for the certification process of such reactors. New accidental transients, following the Fukushima accident learning lesson, will be also studied. The availability of such a complex plant allows performing other exploitation, and studies are foreseen for using it in a wider field of application for integral layout SMR simulation.

An accurate accident analysis requires the measurement of the mixture mass flow rate occurring in a LOCA, when a piping break occurs at high temperature and pressure. For this reason, instruments and methodologies to evaluate mass flow rates, at the break and at other locations of the plant, need to be developed, considering the severe thermal-hydraulic conditions during the blowdown phase (the flow is critical, the average void fraction is higher than 90%, and the flow pattern is annular).

Typically, a set of instruments (Spool Piece) must be installed in order to evaluate the mass flow rate of the two-phases [3–5]. One of such instruments has to be able to measure the void fraction and to identify the flow pattern. In fact, the response of a meter in two-phase flow tends to be highly sensitive to the flow pattern, to the upstream configuration, and to the flow history. As a first step to achieve this purpose, the present study deals with the characterization of a capacitive meter device for annular flow. The behavior of the sensor is tested and advantages and drawbacks are highlighted.

The void fraction in the region of interest is one of the key parameters in gas-liquid two-phase flow systems, as it is used for determining several other important parameters (density and viscosity, velocity of each phase, etc.) and for predicting heat transfer and pressure drops. It can be measured by using a number of techniques, including radiation attenuation (X- or γ-ray or neutron beams) for line or cross section averaged values, optical or ultrasound techniques for local and chordal void fraction measurement, impedance techniques by means of capacitance or conductance sensors for local or volumetric measurements, and quick-closing valves for direct measurement of the volumetric void fraction. The use of the different techniques depends on the applications and whether a volume averaged or a local void fraction measurement is required. All the different techniques are based on the use of a sensor that is sensitive to the variation of the physical properties of the phase mixture and therefore able to detect the presence of one of the phases.

The impedance method is based on the fact that the liquid and gas phases have different electrical conductivity and relative permittivity, and the electrical impedance of a mixture is usually different by the impedance of each component. The gases are generally poor conductors with a low dielectric constant, while the liquids, although not good conductors, assume higher value of the dielectric constant due to a greater concentration of dipoles. In the electrical impedance void meters, the resistive and/or the capacitive response of the two-phase flow region to an electrical field is measured and the resultant resistance and/or capacitance value (impedance) is used to estimate the void fraction. Impedance sensors have been used successfully to measure time and volume averaged void fraction, and their instantaneous output signal has been used to identify the flow pattern in the work of Rochal et al. [6]. The fast response of the impedance meters allows obtaining information about virtually instantaneous void fractions and their distributions across a pipe section. This type of meter can work at high pressure and temperature (in this case the signal is corrected taking into account the effect of pressure and temperature variation on the signals) and also with high flow velocity, and it is more attractive than other techniques from an economic point of view. The measurements of the void fraction with impedance sensors are quasi-local: the sensor determines the percentage of both phases not strictly in a selected cross section of the pipe but in a certain volume, based on the electrodes geometry. The measurements of the impedance take place in a volume defined by the lines of an electric field associated with the electrodes system. The reference control volume for the electrode probe is a function of the electrode surface, of their distance, and of the electrical properties of the medium. Concerning the classical impedance probes, one of the most important drawbacks is the strong sensitivity to the flow pattern that can produce some ambiguity in the signal interpretation; for example, if the sensor produces the same signal for two different flow conditions, a criterion for the flow pattern identification has to be developed.

To address this problem, in the last years, tomography sensors using the impedance probes have been developed and several researches have been carried out [7–9]. The impedance measurements are taken from a multielectrode sensor surrounding a process vessel or a pipeline. The working principle consists of sending a sinusoidal signal to an electrode and measuring the output signal in the remaining electrode. This procedure is repeated for all the other electrodes pairs until a full rotation is completed to get a set of measurements. Each dataset is interpreted by algorithms to compute a cross-sectional electrical capacitance distribution.

The concentration of each phase can be evaluated from the electrical permittivity values of each phase. The advantage is its excellent time resolution arising from the very fast measurement of electrical resistances, while the drawback is the relatively low spatial resolution, since the phases distribution reconstruction is based on measurements at the periphery of the sensor. This problem is stressed in presence of an annular flow as the liquid film at the wall creates a preferential path for the electric field lines that shields the core region and makes the sensor poorly sensitive to the void fraction. The range of interest of the flow parameters, in steady state and transient conditions, has been investigated by means of the RELAP5 code [10] and, for each line of the SPES3 facility, the range of measurement of the different parameters has been identified; the performed simulations show that, in a relative long period of the transients, the two-phase flow corresponds to an annular flow pattern, characterized by very high void fraction and high flow velocity, so that the selected instruments of the Spool Piece must be sensitive to the flow in such conditions. In order to address the drawback, the present study authors have designed and developed a sensor consisting of nine external electrodes and an internal one.

In this paper, the sensor design and the experimental characterization at very high void fraction and in presence of annular air/demineralised-water flow are described. The possibility to obtain information about other thermal-hydraulic parameters is also discussed and an evaluation of the response of the sensor at the variation of the flow pattern has been carried out.

Concerning the annular flow pattern, a large number of experimental and theoretical studies have been performed by several authors (see [11–14]). The flow pattern can be described in a simplified model as a liquid film flowing on the wall of the tube and a gas phase flowing in the center (core flow): some rate of the liquid phase is entrained as small droplets in the gas core. As the liquid flow rate is increased, the droplet concentration in the gas core increases and the droplet coalescence occurs leading to large lumps or streaks as wispy liquid inside the gas core. It is hypothesized that the occurrence of agglomeration is due to a fundamental instability of the gas-droplet core flow: if positive wave growth coefficients are present, the system is classified as unstable and belongs to the wispy-annular flow regime [12], while fully stable systems are classified as belonging to the annular flow regime.

The annular flow is quite difficult to be observed using ordinary visualization techniques since the liquid film in this regime tends to screen the core region. The present study investigates also the use of the ECP sensor to detect the phase distribution inside the core region of the annular flow.

ELECTRICAL CAPACITANCE PROBE (ECP)

The sensor ECP has been developed by the SIET Company and consists of 10 electrodes: 9 external and one central (Figure 1). The inner and the outer diameters of the Plexiglas pipe, where the probe is mounted, are 80 mm and 90 mm. The external electrodes (steel stripes of 400 mm length and 5 mm width) are spaced with 22.5° angle only on a half-circumference of the pipe, due to the vertical flow symmetry. The angle corresponds to an external chord of 17.56 mm and an internal chord of 15.6 mm. The sensor geometry has been developed in half of the pipe circumference as the symmetry of the vertical flow has been verified. The main purpose of the sensor is the evaluation of the effect of the angular electrode distance on the output signal.

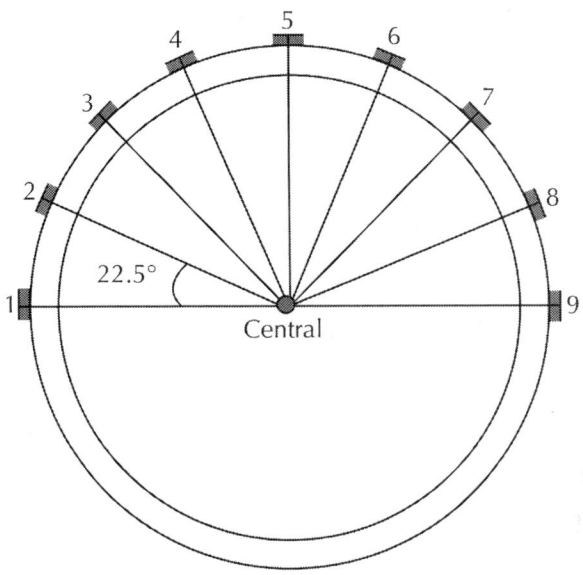

Figure 1: Schematic of the ECP.

The external electrodes are pasted on the Plexiglas pipe and are welded with the conductor that allows the link with the electronic part, while the internal electrode is connected through a metallic support at the outside of the pipe.

The electrodes are connected in an electronic circuit by several reed relays and two insulation transformers in order to prevent common mode disturbances. The reed relays technology has been selected for its simplicity and its features: very small closing and opening time, very high resistance when open, very small resistance when closed, very small electrical capacity, and long life.

Each external electrode is connected, at the upper and lower extremity, to two reed relays to activate, in a predefined sequence, the excited electrode and the measuring one; the internal one is connected only in the upper extremity and it is always used as a measuring electrode, when the corresponding reed relay is activated.

EXPERIMENTAL FACILITY AND TEST SECTION

The experimental facility consists of the feed water and the feed air loops equipped with instruments to measure the single-phase flow parameters (flow meter, temperature, and pressure). The water mass flow rate is measured by a rotameter, while the air flow rate is measured by a calibrated orifice flow meter, whose accuracy is 2% full scale value. The facility, which is shown in Figure 2, consists of a vertical Plexiglas tube (80 mm inside diameter and 4 m long). The test section is transparent (Plexiglas) in order to visualize the flow pattern. The air and the water are injected inside the pipe at the bottom of the tube and mix at about 400 mm from the test section inlet (Figure2); the water is injected through a porous bronze. The 2.5 m long test section is equipped with two pneumatic quick-closing valves (QCVs) to measure the volumetric void fraction; the uncertainty associated with the void fraction measurement has been estimated as $\Delta\alpha = \pm0.0012$.

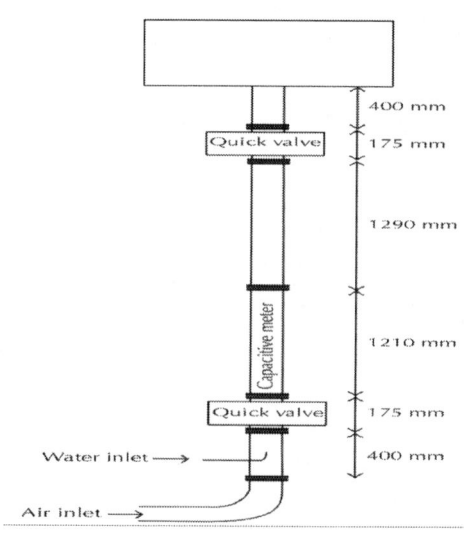

(a)

(b)

Figure 2: Experimental facility: test section schematic (a) and picture (b).

Downstream of the upper valve the air and the water are separated in a tank at atmospheric pressure. Experiments are carried out at a constant water temperature of 20°C, and the absolute pressure is measured at the inlet of the test section by a pressure transducer Rosemount 3051/1.

The presence of the internal electrode (whose length is 400 mm and diameter is 5 mm) in the center of the pipe affects the mechanisms of the dispersed flow; a liquid film has been observed on it. Okawa et al. [15] have shown in their experiments that the deposition rate of droplets was markedly increased if the present flow obstacle was placed inside the flow channel and the deposition rate was approximately 1.5 times larger than the deposition rate with no obstacle. The presence of the central obstacle together with a higher diameter and the coaxial liquid injection can affect the waves in the liquid film and the droplet coalescence in the core region also at rather low liquid superficial velocities.

EXPERIMENTAL METHODOLOGY AND SIGNAL ACQUISITION

The signals from the sensor are acquired by using the NI USB-6259 DAQ (Data AcQuisition) and are managed using a LabVIEW program. The predefined measurement sequence is read and the corresponding reed relay is activated by using a 5 V DC signal. The excitation signal is sent to the electrodes (sinusoidal signal with an excitation frequency f_{ex} = 25 kHz and an amplitude of 5 V) and the output signal is sampled using a frequency f_{ac} of 250 kHz; the (RMS root mean square) value corresponding to 2000 samples is acquired. The excitation frequency has been chosen considering that the experiments are carried out with demineralised water; in this case, the sensor output signal is directly proportional to the fluid capacitance.

The measurement sequence is defined as follows: the external electrodes are excited in sequence and for each one the output signal of the other external electrodes are read; after that scan the output signal relative to the central electrode is acquired for each excited external electrode. For every couple of electrodes, the sampling time of RMS value is about 33 ms so that the total scanning time depends on the selected measurement sequence. Before each set of experimental runs, the static values of the signals for air and water are measured. Then, for each run, the mass flow rates of water and air are fixed and, when the flow is in steady state, the pressure value and the signals from the ECP sensor are acquired. At the end of the measurement sequence, the volume averaged void fraction is measured by means of the quick-closing valves technique.

EXPERIMENTAL DATA RANGE

In order to evaluate the response of the sensor at high void fractions, corresponding to annular flow, a mixture of air and demineralized water is introduced in the test section. The mass flow rate ranges from 0.094 to 0.15 kg/s for air and from 0.002 to 0.021 kg/s for water, corresponding to a void fraction higher than 95%. The tests have been performed at atmospheric pressure. The volumetric void fraction (measured with QCV technique) increases with gas superficial velocity J_g and decreases with J_l, as shown in Figure 3.

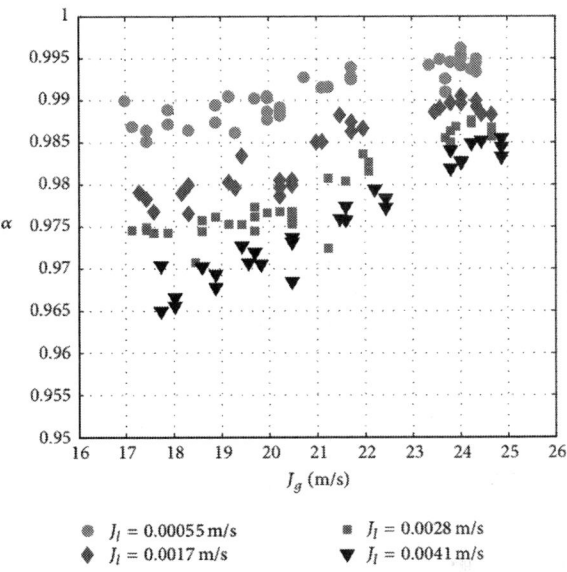

Figure 3: Experimental volumetric void fraction as a function of water and air superficial velocities.

Figure 4 shows the volumetric void fraction as a function of the flow quality: a small flow quality variation at constant J_l corresponds to large void fraction variations, while the slope of the void fraction versus x increases as J_l decreases.

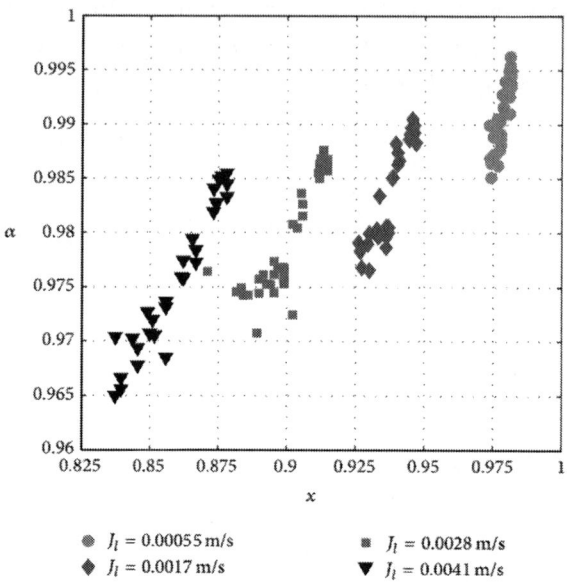

Figure 4: Experimental volumetric void fraction as a function of the flow quality at different water superficial velocity.

CAPACITANCE PROBE EXPERIMENTAL RESULTS

The first ECP qualification step is the characterization of the sensor for single-phase conditions; so, before each test, measurements with air and demineralized water have been carried out in order to evaluate the signal noise level and to normalize the signals for two-phase flows; the single-phase measurements are used to determine the electrodes shape function of the sensor. In Figure 5, the RMS signals measured between electrode 1 and the other external electrodes are presented, as a function of the value of the angle θ between the electrodes. The measured RMS value is proportional to the electrical capacitance between the measuring electrodes; it also depends on the excitation frequency and on the resistance that is different between air, water, and two-phase mixture, also with demineralized water. As shown in Figure 5, the signal variation range is very limited due to the presence of the quite large Plexiglas thickness (5 mm) and to the absence of any signal amplification.

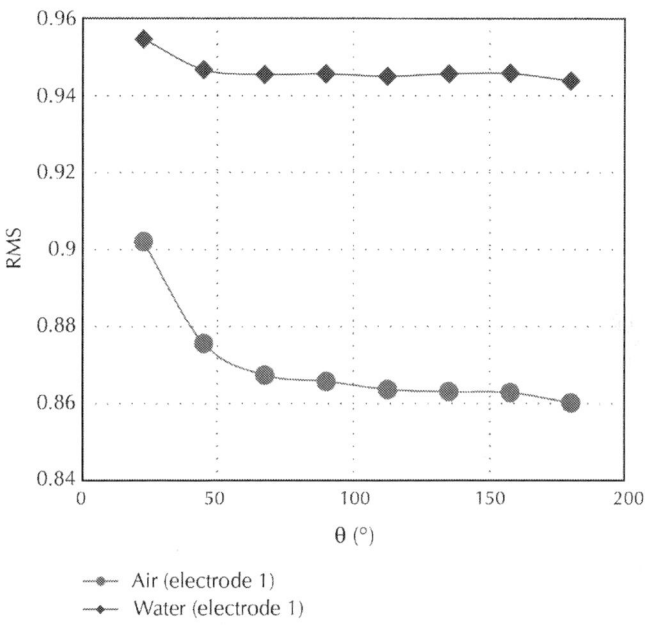

Air (electrode 1)
Water (electrode 1)

Figure 5: Single-phase ECP signals as a function of the angle θ for the excited electrode 1.

Moreover, the theoretical ratio between the water electrical permittivity and the air electrical permittivity, which equal to 80, is reduced, in the practical case, at a value from 1.05 to 1.1 depending on the electrodes distance. The ratio is lower at $\theta = 22.5°$ due to the strong influence of the wall. The angle dependence is higher for the signal measured in air flow, while it seems limited for single-phase water flow

For each electrodes combination, the ratio RMS_g/RMS_l shows a rather high repeatability. The ratio of the two-phase RMSs, obtained by the measurements between the external electrodes, as a function of the angle θ is shown in Figure 6

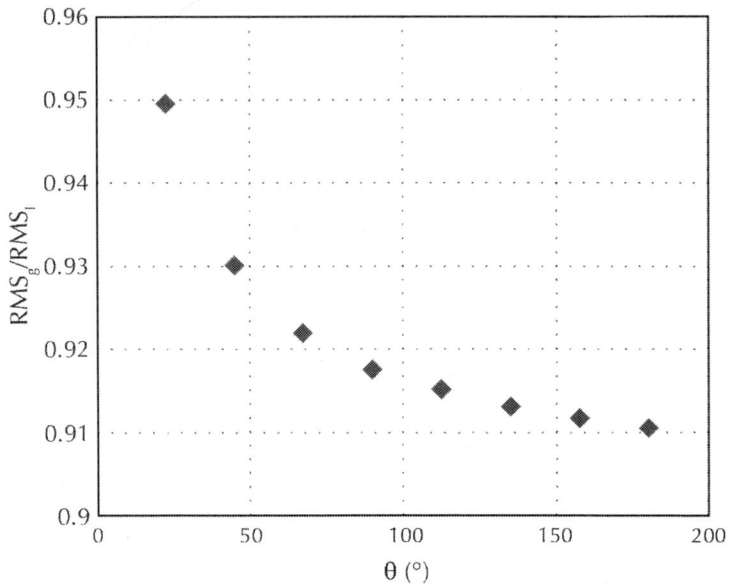

Figure 6: Single-phase ECP signals ratio as a function of the angle θ, for the excited electrode 1.

The analysis of the signals of the central electrode shows a higher sensitivity due to the direct contact of the probe with the mixture. In this case, the mean RMS_g/RMS_l ratio is equal to 0.582 ± 0.002

In order to take into account the single-phase signal variations, the two-phase flow measured values are normalized as follows:

$$V_{ij}^* = \frac{RMS_{TP\text{-}ij} - RMS_{l\text{-}ij}}{RMS_{g\text{-}ij} - RMS_{l\text{-}ij}},$$

(1)

where the subscript ij identifies the measuring electrodes combination.

Because of the physical differences between the internal and external electrode response, the signal of external and internal electrodes has been analyzed separately.

External Electrodes Signals

In order to evaluate the behavior of the sensors as a function of the mean void fraction, only the $i\text{-}j$ combination signals are considered (deriving the mean value and the standard deviations); while, in order to evaluate the interface phenomena in the liquid film and in the core region, the time dependent signals (an example is reported in Figure 7 for the electrodes combination 1-2 and test volumetric void fraction α equal to 0.995 (a) and 0.977 (b)) have to be considered. The signals evolution is shown in Figure 7: the single-phase signal is very stable in time so that the two-phase flow signals fluctuation can be related with the dynamic evolution of the flow; the third digit of the RMS has to be considered in order to take into account the film and the core flow dynamics. In Figure 7 the signal has been compared with water or air depending on the nearest mean RMS value.

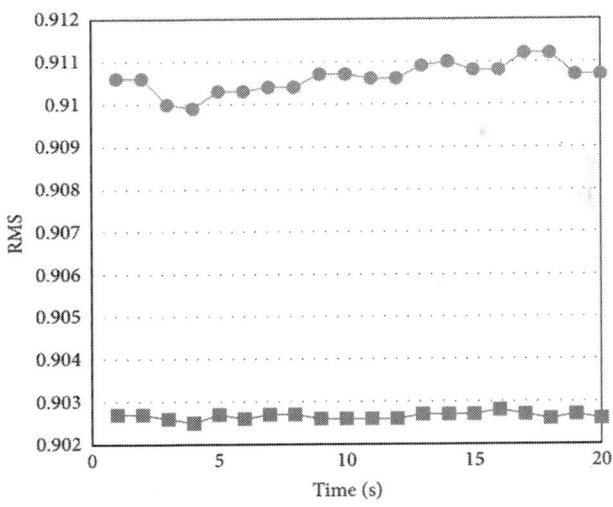

Electrodes combination: 1-2
—■— Air
—●— α = 0.995

(a)

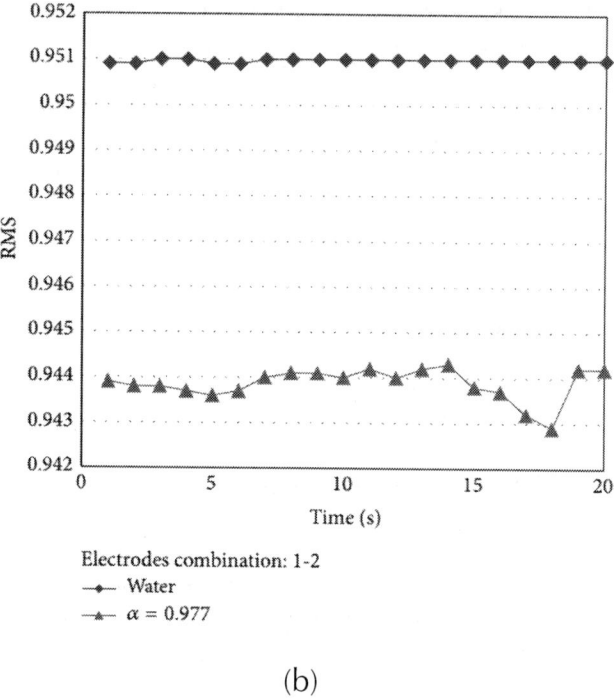

(b)

Figure 7: RMS time evolution for 1-2 electrodes combination compared with the single-phase RMS time evolution: (a) $\alpha = 0.995$; (b) $\alpha = 0.977$.

Each combination is analyzed in terms of dependence on the fluid-dynamic quantities (experimental void fraction, superficial velocities of air and water) and geometry parameters (θ). A typical set of curves is reported in Figures 8 and 9. The sensor is sensitive to very little variation of the void fraction and this variation is detected from all the measuring electrodes combinations. As reported in the following pictures, the normalized signal v_{ij}^{*} is a function of angle θ, liquid film thickness δ, and volumetric void fraction α (measured by the QCV technique).

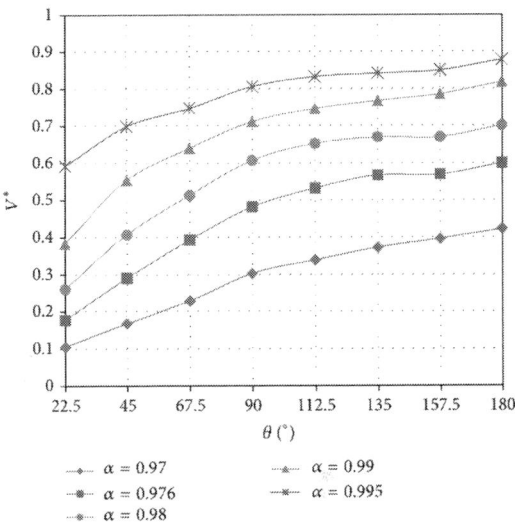

Figure 8: Normalized signal v_{ij}^* for $i=1$ and $j=2–9$ as a function of angle θ for different experimental volumetric void fractions α.

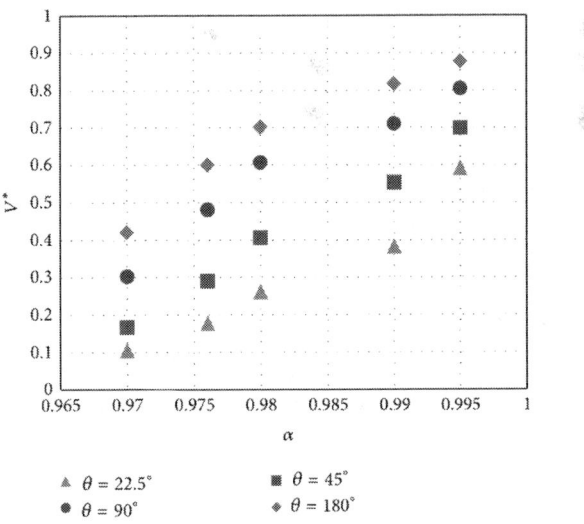

Figure 9: Normalized signal v_{ij}^* as a function of the experimental volumetric void fraction α for different angles θ.

The relationship between the capacitance and the void fraction depends not only on the dielectric values of the two-phases, but also on the surfaces of the sensors, on the separation distance between the two electrodes, and on the voltage distribution inside the measurement volume, which in turn depends on the phases distribution.

As regards the flow pattern, it is possible to develop a qualitative model for the signal variation: in the annular flow, the measurements taken from the external electrodes are not very sensitive to the core region flow, because the preferential path of the electrical field lines is located in the continuous liquid film. The analysis of the signal shows that the measurement, taken between close electrodes ($\theta = 22.5°$), is more sensitive to the presence of the liquid film compared to the electrodes having higher distances, whose signal is affected by the flow distribution in the core region. In the hypothesis of axial symmetric flow, and in order to evaluate the sensor sensitivity to the mean void fraction variation, all the signals measured between electrodes, placed at the same angular distance, are used to evaluate the average normalized signal at the angle θ. The average signals measured between electrodes at 22.5°, 45°, 90°, and 180° are presented in Figure 10 as a function of the

volumetric void fraction measured by the QCV technique. v_{ij}^* (1) increases with θ at constant void fraction and approaches the air value at higher angular distance. No significant difference has been found between the signals of electrodes at 90° and 132.5° and at 157.5° and 180°.

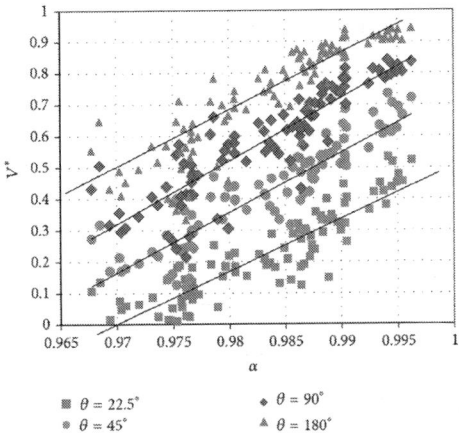

Figure 10: Mean signal measured in the external electrodes as a function of the experimental volumetric void fraction α.

In Figure 11, the best-fit coefficients (*a*: slope, *b*: intercept) of the curves reported in Figure 10 are shown. The curves are parallel for angles higher than 45° while for the angles equal to 22.5° and 45° the signal dependence is different, due to the larger effect of the Plexiglas wall and of the liquid film thickness.

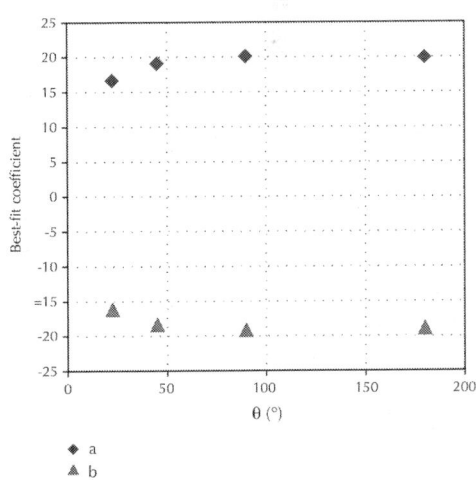

Figure 11: Best-fit coefficient dependence on the measuring angle θ for the curves shown in Figure 10; *a*: slope, *b*: intercept.

Central Electrode Signals

The signal of the central electrode, which is in direct contact with the fluid, is sensitive to the mean cross section void fraction and could be directly related to the measured void fraction and to the amount of liquid droplets in the core region.

In Figure 12, the mean signal measured between the central electrode ($j = 10$) and the external electrodes ($i = 1 : 9$) is represented as a function of the experimental void fraction, measured by means of the QCV technique: the signal depends linearly on the void fraction at values higher than 0.98 and it is characterized by a higher standard deviation at lower values. In the tested range, the observed flow pattern is annular, with a rather low turbulence at the film interface, for liquid superficial velocities lower than 0.00152 m/s, while, at higher water mass flow rates and void fraction values lower than 0.98, the flow pattern tends to become more turbulent with a greater mass of liquid that is entrained inside the core region as previously discussed.

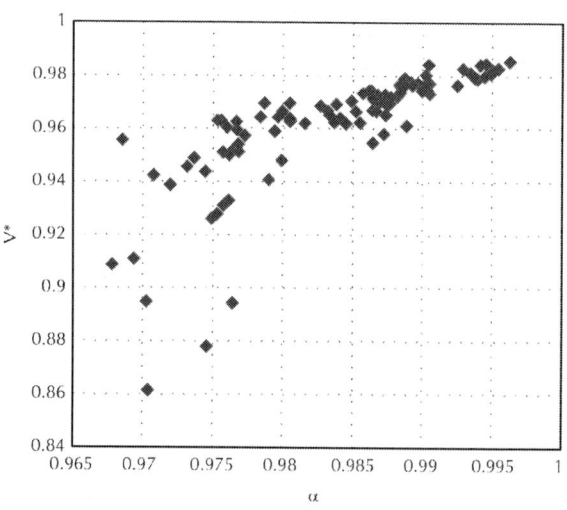

Figure 12: Mean signal measured in the central electrode as a function of the experimental volumetric void fraction α.

The annular flow pattern is more regular at higher experimental void fraction as it is shown in Figure 12.

Annular Flow Model and Signals Interpretation

Due to the symmetry of the flow, the annular flow regime in a vertical channel can be analyzed in a simplified scheme as a liquid film region and a core region. The liquid film is characterized in terms of film thickness, frequency, and amplitudes of the waves at the liquid-gas interface, while the core region is characterized in terms of the mean void fraction value. By increasing the air flow rate, the liquid entrainment from the film to the gas core increases, while by increasing the water flow rate the amplitude of the waves increases and the flow tends to become more turbulent with a higher fraction of the water phase that is entrained in the core region and with an irregular wavy film flow. In order to analyze the experimental results, some relevant flow parameters have been expressed by means of a simplified model. The parameters evaluated with the model are then used to interpret the sensor signals. In an annular flow, the void fraction can be written as follows:

$$\alpha = \alpha_c \cdot (1 - \alpha_d),$$

(2)

Where

$$\alpha_d = \frac{A_d}{A_c},$$

$$\alpha_c = \frac{A_c}{A} = 1 - \alpha_f,$$

(3)

where A_c is the core region area, α_d is the cross section fraction occupied by the droplets in the core, α_c is the fraction of the pipe cross section occupied by the core region, and α_f is the fraction of the pipe cross section occupied by the film region. Then, the value of α_c is an indirect index of the liquid film thickness.

Ishii and Mishima [13] derived the criteria for the onset of entrainment based on the balance of the forces acting on the waves to characterize the interface evolution in annular flow. The fraction of the liquid flux flowing as droplets (E_∞) is derived by Ishii and Mishima [13] and Hazuku et al. [14] as follows:

$$E_\infty = \tanh\left(7.25 \cdot 10^{-7} \cdot \text{We}^{1.25} \cdot \text{Re}_f^{0.25}\right),$$

(4)

where the Weber number is

$$\text{We} = \frac{\rho_g \cdot J_g^2 \cdot D}{\sigma}\left(\frac{\rho_l - \rho_g}{\rho_g}\right)$$

(5)

and the Reynolds number is

$$\text{Re} = \frac{\rho_l \cdot J_l \cdot D}{\mu_l},$$

(6)

where ρ_g, μ_g, J_g and ρ_l, μ_l, J_l are respectively the density, the dynamic viscosity, and the superficial velocity of air (subscript g) and water (subscript l), while σ is the surface tension.

In the present tests, the liquid Reynolds number Re_l ranges from 44 to 332 and the Weber number We ranges from 2200 to 7000.

In a pure annular flow, the liquid flows only in the film region, so the maximum value of liquid film thickness δ_{max} is evaluated from the measured void fraction as follows:

$$\alpha = \frac{(D - 2 \cdot \delta_{max})^2}{D^2}.$$

(7)

Under the hypothesis of pure annular flow, the liquid film velocity U_l and the core gas velocity U_g are evaluated as follows:

$$U_g = \frac{J_g}{\alpha},$$

$$U_l = \frac{J_l}{1 - \alpha}.$$

(8)

In Figure 13, the liquid film thickness is shown as a function of the superficial velocity of the two-phases. Actually this model is valid for the pure annular flow pattern, but, by introducing the correction due to the entrainment (Figure 14), the core gas fraction is evaluated more correctly. The velocities of the phases are corrected assuming the same liquid droplets velocity as that of the gas in the core region:

$$U_g = \frac{J_g}{(\alpha_c - \alpha_d)},$$

(9)

$$U_l = (1 - E_\infty) \cdot U_f + E_\infty \cdot U_g,$$

(10)

$$U_f = (1 - E_\infty) \cdot \frac{J_l}{(1 - \alpha_c)}.$$

(11)

The velocity of the two-phases is shown in Figures 15 and 16.

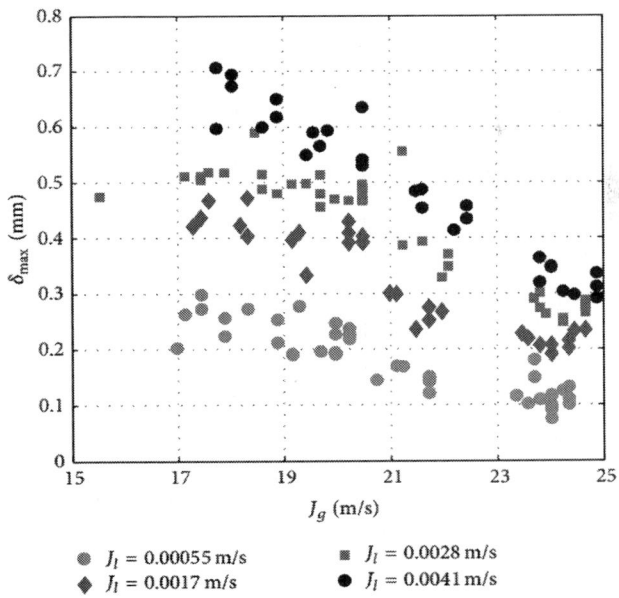

Figure 13: Maximum film thickness (7) as a function of the experimental water and air superficial velocity.

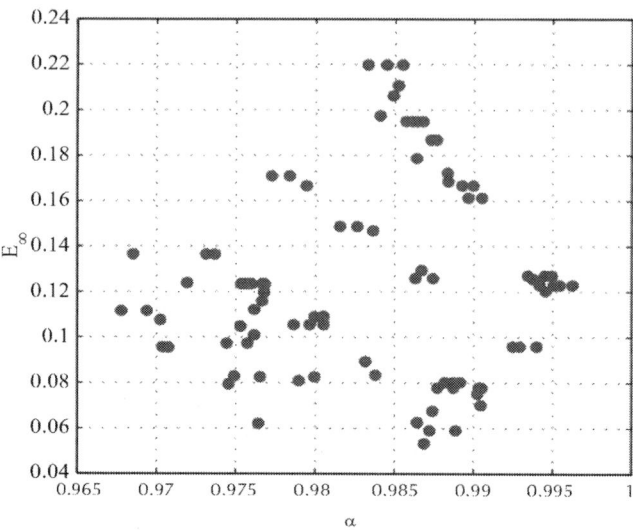

Figure 14: Equilibrium entrainment (4) as a function of the experimental volumetric void fraction α.

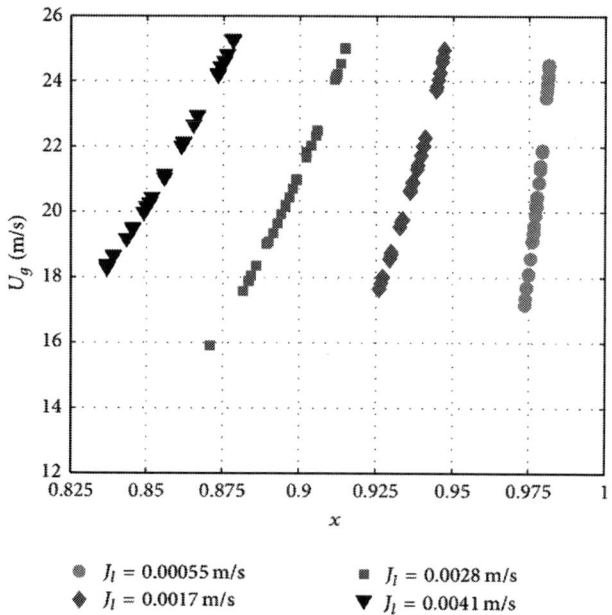

Figure 15: Air velocity as a function of the experimental flow quality (9).

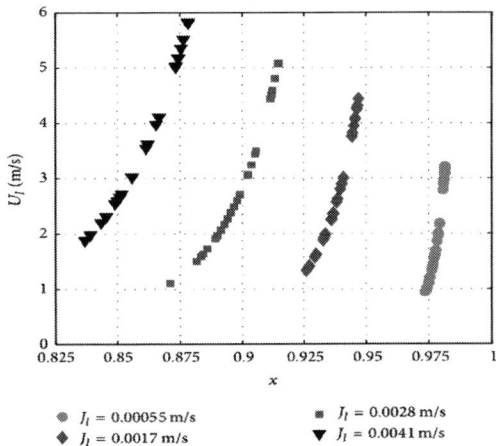

Figure 16: Water velocity as a function of the experimental flow quality (10).

In Figure 17, the derived slip ratio ($S=U_g/U_l$) is shown as a function of the flow quality, at different liquid superficial velocity.

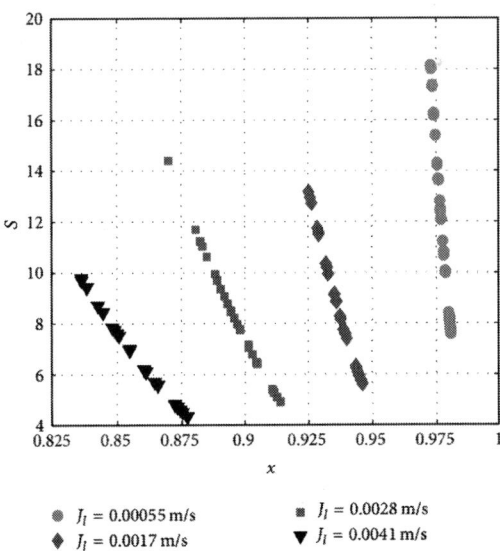

Figure 17: Slip ratio $S=U_g/U_l$ (evaluated on the basis of Ishii's model) as a function of the experimental flow quality.

The liquid film thickness is corrected, as shown in Figure 18, taking into account the amount of liquid that is entrained into the core region:

$$\alpha_c = \frac{(D - 2 \cdot \delta)^2}{D^2}.$$

(12)

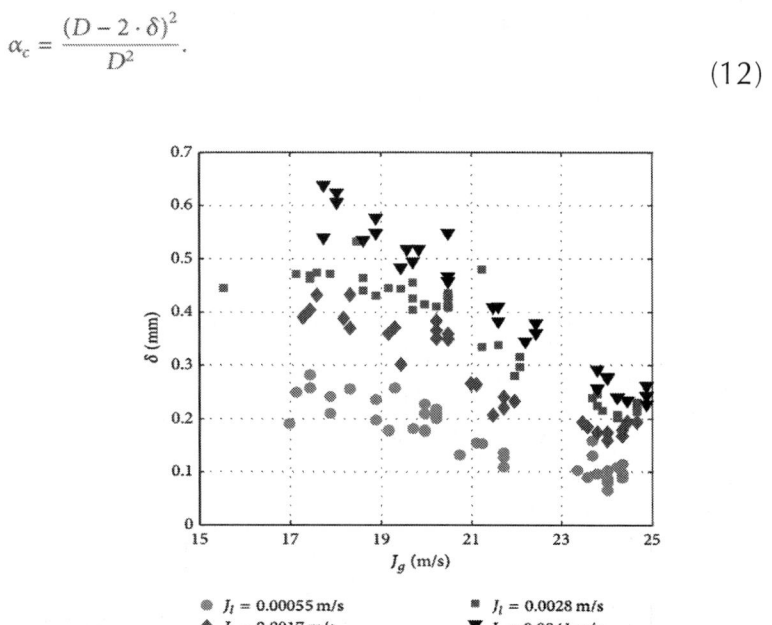

Figure 18: Corrected mean film thickness as a function of the experimental water and air superficial velocities.

In order to analyze the dependence of the sensor signals on the measured void fraction, the data of the different electrodes combination are related to the liquid film thickness δ by means of the core void fraction α_c and to the liquid fraction in the core region by means of the droplets fraction α_d. A model correlating the probe signals to the two-phase flow parameters, as α_c, α_d, J_l, J_g, pressure p, geometry, and fluid parameters is required. For an axial symmetrical two phase flow at constant pressure and temperature, a simple model with two parameters and two signals (from an external electrode and from the central one) can estimate the film thickness, the liquid core droplet fraction, and the void fraction. In Figure 19, the signals measured in the central electrode are shown as a function of the core fraction. The flow pattern characteristics, from rather regular to more disturbed annular flow, are clearly detected. In Figure 20, the same signal is evaluated as a function of the droplets fraction: the internal probe is more sensitive than the external ones to the effect of the liquid and gas velocity change.

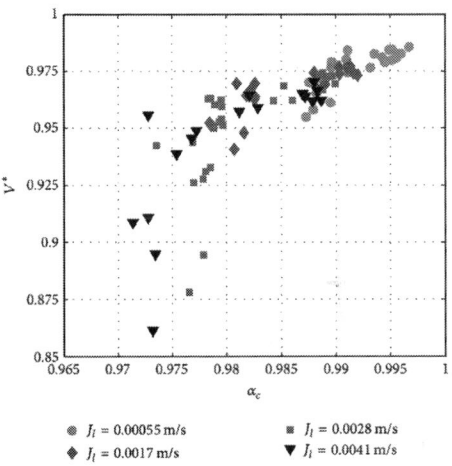

Figure 19: Normalized average signals of the central electrode, as a function of the core void fraction (evaluated on the basis of Ishii's model).

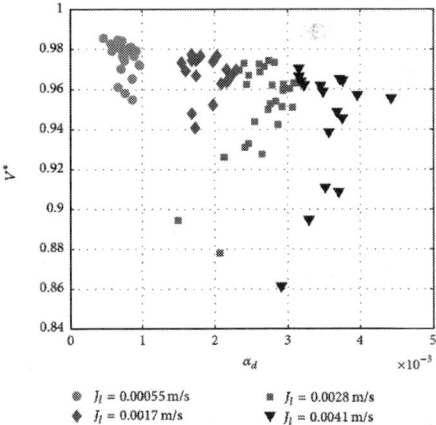

Figure 20: Normalized average signals of the central electrode, as a function of the core droplets fraction (evaluated on the basis of Ishii's model) at different experimental liquid superficial velocities.

In the future, the direct measurement of the local liquid film thickness and of the frequency of the film waves will be performed and their influence will be considered in the signal analysis.

Figure 21 shows the evolution of the signals with the film thickness, evaluated using the Ishii model, for the different measuring angles. The dependence is linear for all the angles, but the intercept coefficient (b) and the slop (a) of the best-fit curves depend on the angle θ, as reported in Figure 22.

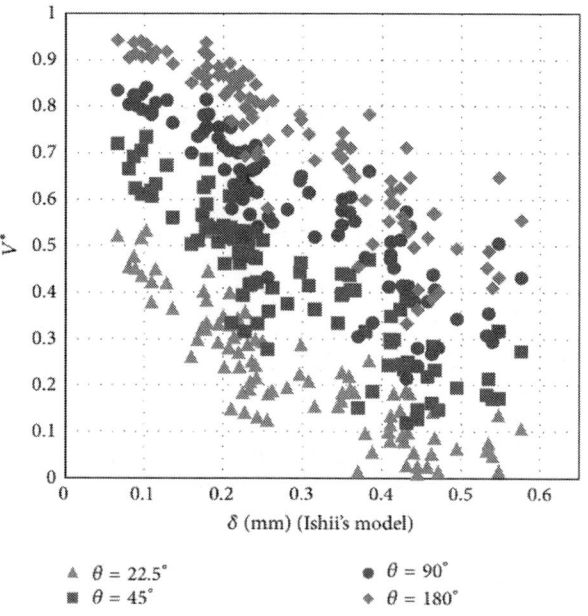

Figure 21: Average signal measured in the external electrodes as a function of the liquid film thickness (evaluated on the basis of Ishii's model).

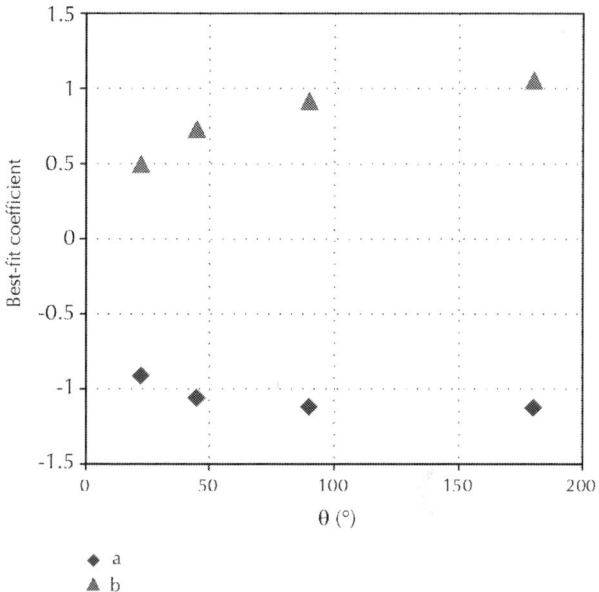

Figure 22: Best-fit coefficient dependence on the measuring angle for the curves shown in Figure 21; *a*: slope, *b*: intercept.

CONCLUSIONS

The paper presents the characterization of an electrical capacitance probe (ECP) that has been developed by the SIET Company and that consists of ten measurement electrodes, nine external and an internal one. The presence of the internal electrode introduces a little flow disturbance but allows one to obtain important information on the phase distribution inside the core region of the annular flow.

The response of the sensor has been characterized in terms of single-phase flow sensitivity and signal variation dependence; geometry and fluid-dynamic influences on the signal have been investigated. Although the sensitivity in single-phase flow is low, two-phase flow void fraction variations lower than 1% have been detected. The presence of the central electrode allows the evaluation of the average cross section void fraction also in annular flow, where the liquid film is the principal cause of the sensor low sensitivity. The signal measured in the central electrode is linear with the void fraction for values higher than 98% and it is quite sensitive to the flow pattern: the flow pattern, from rather regular to more disturbed annular flow, has been clearly characterized. The variation of the signal measured in the external electrodes has been related to the average liquid film thickness that is evaluated by means of a simple model. The tests have shown the potentiality of this technology for the measurement of two-phase flow parameters at very high void fraction conditions (higher than 95%).

On the ground of the present experimental results, the local film thickness and the frequencies of the film waves will be investigated in order to develop a model of the sensor suitable to evaluate the chordal void fractions and other fluid-dynamic parameters, also for a higher measurement parameters range. In this context, the time analysis of the electrodes signals can be carried out to identify such important fluid-dynamic parameters as well as the film interface oscillations.

In order to evaluate the performance of the sensor in the entire range of the flow patterns that are expected during the transient accident at the SPES3 facility, according to RELAP5 simulation, new experimental tests are currently underway, while the SIET Company is developing a sensor, having the same geometry of the present ECP, which is able to operate at the real break line conditions (200°C and 15 bar) [16]. In order to operate at high temperature and pressure, the Plexiglas pipe, in which the electrodes are located, has been substituted with a Pyrex pipe, and for safety purpose the sensor has been located inside a stainless steel pipe that can be pressurized from outside; the new probe has been thermally and mechanically tested and it is now ready for thermal-hydraulic tests. Moreover, in order to evaluate the mass flow rate of the phases during a LOCA, the experimental characterization of a Spool Piece, including the ECP, is underway at Politecnico di Torino.

ACKNOWLEDGMENTS

The present research has been supported by ENEA and by the Ministry of Economic Development. The authors wish to thank R. Costantino and G. Vannelli for their technical support.

REFERENCES

1.   M. Carelli, L. Conway, M. Dzodzo, et al., "The SPES3 experimental facility design for the IRIS reactor simulation," Science and Technology of Nuclear Installations, vol. 2009, Article ID 579430, 12 pages, 2009. ·

2.   R. Ferri, A. Achilli, C. Congiu, et al., "SPES3 facility and IRIS reactor numerical simulations for the SPES3 final design," in Proceedings of the European Nuclear Conference (ENC ‹10), Barcelona, Spain, May 2010.

3.   M. Greco, R. Ferri, A. Achilli et al., "Two-phase flow measurement studies for the SPES3 integral test facility for IRIS reactor simulation," in Proceedings of The 18th International Conference on Nuclear Engineering (ICONE ‹18), pp. 305–316, Xi›an, China, May 2010.

4.   M. De Salve, G. Monni, and B. Panella, "State of art and selection of techniques in multiphase flow measurement," Report RdS/2010/67, ENEA, 2010.

5.   C. Bertani, M. De Salve, M. Malandrone, G. Monni, B. Panella, and A. Mosetto, "SPES-3 facility analysis," Report RdS/2010/68, ENEA, reference data for postulated accident simulation; criteria for general and special instrumentation selection, 2010.

6.   M. S. Rochal, E. L. L. Cabral, and J. R. Simões-Moreira, "Capacitance sensor for void fraction measurement in a natural circulation refrigeration circuit," in Proceedings of the International Nuclear Atlantic Conference (INAC ‹09), 2009.

7. Z. Huang, B. Wang, and H. Li, "Application of electrical capacitance tomography to the void fraction measurement of two-phase flow," IEEE Transactions on Instrumentation and Measurement, vol. 52, no. 1, pp. 7–12, 2003.

8. Y. Wu, H. Li, M. Wang, and R. A. Williams, "Characterization of air-water two-phase vertical flow by using electrical resistance imaging," Canadian Journal of Chemical Engineering, vol. 83, no. 1, pp. 37–41, 2005.

9. W. Warsito and L. S. Fan, "Measurement of real-time flow structures in gas-liquid and gas-liquid-solid flow systems using electrical capacitance tomography (ECT)," Chemical Engineering Science, vol. 56, no. 21-22, pp. 6455–6462, 2001.

10. R. Ferri and C. Congiu, "SPES3-IRIS facility RELAP5 base case transient analyses for design support," SIET Document 01 489 RT 09 Rev.0, 2009.

11. A. W. Bennett, G. F. Hewitt, H. A. Kearsey, R. K. F. Keeys, and M. P. C. Lacey, "Flow visualisation studies of boiling at high pressure," Proceedings of the Institution of Mechanical Engineers, vol. 180, part 3C, pp. 1–11, 1965.

12. G. F. Hewitt and D. N. Roberts, "Studies of two-phase flow patterns by simultaneous X-ray and flash photography," UKAEA Report AERE-M 2159, London, UK, 1969.

13. M. Ishii and K. Mishima, "Two-fluid model and hydrodynamic constitutive relations," Nuclear Engineering and Design, vol. 82, no. 2-3, pp. 107–126, 1984.

14. T. Hazuku, T. Takamasa, T. Hibiki, and M. Ishii, "Interfacial area concentration in annular two-phase flow," International Journal of Heat and Mass Transfer, vol. 50, no. 15-16, pp. 2986–2995, 2007.

15. T. Okawa, A. Kotani, N. Shimada, and I. Kataoka, "Effects of a flow obstacle on the deposition rate of droplets in annular two-phase flow," Journal of Nuclear Science and Technology, vol. 41, no. 9, pp. 871–879, 2004.

16. C. Randaccio, "Prove a caldo di una sonda capacitiva per la misura del grado di vuoto in miscela bifase,"SIET Document 01 876 RP 12, 2012.

Characterization of two Phase Flows in Chemical Engineering Reactors

S.L. Kiambi[a], H.K. Kiriamiti[b], and A. Kumar[b]

[a]School of Chemical Engineering, University of KwaZulu Natal, Durban, South Africa
[b]Department of Chemical Engineering, Moi University, Eldoret, Kenya

ABSTRACT

Most industrial processes like fermentation, hydrogenation, oxidation, water treatment, petrochemical, nuclear and aerospace involve intimate contact between continuous phase and dispersed phase. Bubble columns and external loop airlifts are commonly used in these operations. Although these reactors are widely used, and extensive research has been carried out there exists no perfect model to characterize the local hydrodynamics and mass transfer. Computational fluid dynamics has also evolved recently trying to model the flow and transfer within these reactors but a lot of results are conflicting. Besides, there is a need to validate these results with experimental work. This work is dedicated to the experimental methods of measuring the local parameters such as the gas hold-up, the bubble velocities, the liquid velocity, bubble sizes. Two intrusive methods are used in this study, hot film anemometry for measurements of liquid phase and bi-optic probe for the gas phase.

Although intrusive methods may interfere with fluid flow, presently they are the most adapted for real industrial processes with opaque equipment walls and high gas hold-ups compared to non-intrusive methods such as imagery. In this study, an external loop airlift of 16.5 l capacity is used. Local variables are measured in the riser of an external loop airlift reactor in air/water medium. The results are presented in the form of the radial profiles of void fraction, bubble diameter, liquid velocities at superficial gas velocities ranging from 0.03 to 0.11 m/s. The axial variations of the same parameters are also investigated. The results suggest that the bi-optical probe and hot film anemometer can reliably predict flow characteristics in high gas hold up contactors.

GRAPHICAL ABSTRACT

▶ Local hydrodynamics in an external loop airlift reactor are characterized in an air–water system. ▶ Double optical fiber is used for gas characteristics in the riser. ▶ Hot film anemometry is used to characterize the liquid phase. ▶ The results obtained are in agreement to a number studies found in the literature. ▶ CFD simulations could also be done to validate the data obtained. The results are presented in form of radial profiles across the riser's diameter.

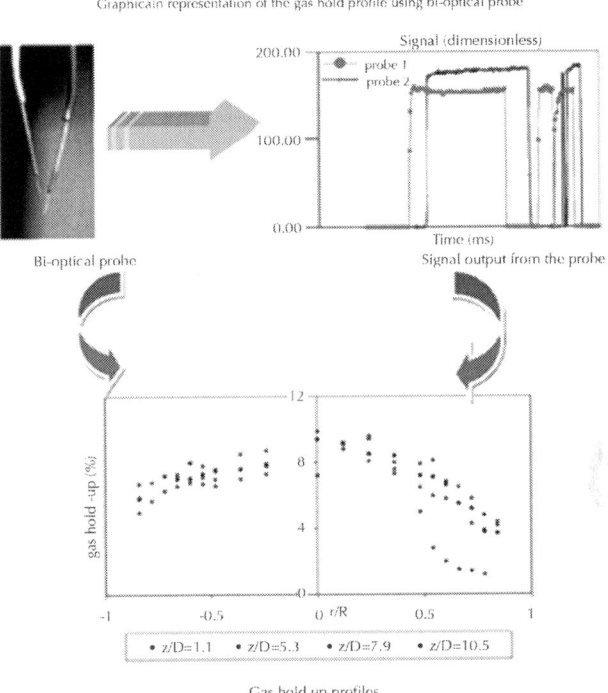

Graphicaln representation of the gas hold profile using bi-optical probe

Gas hold up profiles

INTRODUCTION

An accurate local measurement of two-phase flow characteristics is important for the development of two-phase flow models. Computational fluid dynamics (CFD) is developing rapidly in chemical engineering and models arising from these simulations need accurate experimental data to be validated. In gas–liquid contactors, a precise knowledge of interfacial area is essential for quantification of mass transfer efficiency. Multi-sensor needle probes are a very common tool for gas–liquid local hydrodynamics investigations in high void fraction bubbly flows.

From the bi-optical probe, local two-phase flow parameters such as bubble frequency, bubble velocity, void fraction, bubble chord lengths, bubble size and interfacial area concentration can be deduced. The principle of this instrument, in which an individual probe contains an optical fiber, is described and discussed in [1], [2] and [3]. Changes in the intensity of light reflected from the immersed end of the fiber by the passage of an air/water interface are detected by an optical amplifier at the other end. Using this technique with a double probe, we made measurements in for gas flows rates of 0.03–0.11 m/s in an external loop airlift reactor. Furthermore, using the hot film anemometer the characteristics of the liquid phase can also be measured. The objective of this study is to provide a data base of experimental results in an external loop airlift reactor. This data could be very useful for CFD modeling.

MEASUREMENT TECHNIQUES

Bi-Optical Probe

The bi-optical probe and the opto-electronic equipment used in this study were fabricated by RBI (France). The two glass fibers are 3.1 mm apart and the diameter of the tip is 40 μm as shown in Fig. 1. The phase differentiation is based on the refractive indices between air (n_{air}=1), the liquid water (n_{H2O}=1.33) and the fiber (n_{fibre}=1.15). According to Snell's law the total angle of reflection of light is greater in case of liquid than for the gas. Hence when the probe is in contact with the liquid the light is refracted and lost in the medium while when in contact with the gas, all the light is reflected back into the opto-electric cell [4]. The two fibers are connected to opto-electronic cell which emits the light signal and transforms the feedback light signal to an electric signal on returning to the equipment. The response time of the cell is very short (0.5–1 μs) thus local phase changes are detected instantaneously. The opto-electronic cell is adjusted in such a way the output signal is between 0, when in liquid phase and 5V for the gas phase. Data acquisition is done by a computer through a conversion card from analog to digital. The sampling frequency is 6.25 kHz which is sufficient to acquire representative signals. This frequency allows the acquisition to be done over 20 s. The signal is then digitized and threshold is carefully chosen to distinguish between the bubble and the liquid signal.

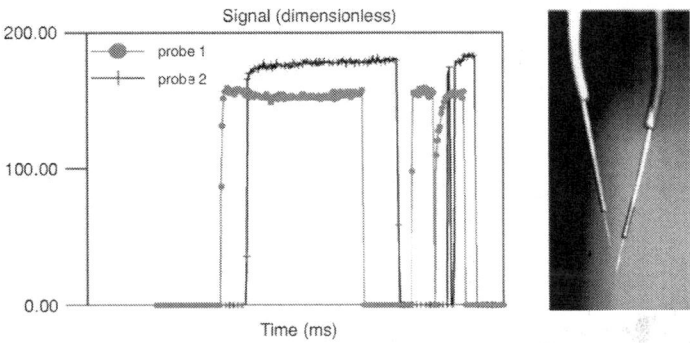

Figure 1: Bi-optic fiber probe and an output signal.

The bubble frequency is measured by counting the number of bubbles pierced by the first probe in a given duration [5] and [6]. Gas hold-up is obtained by summing the bubble signal part divided by the duration considered. $\varepsilon = \dfrac{\sum_i t_i}{t_o}$. Where t_i, represents the transit time of ith bubble on the probe, and t_o total time duration. For every bubble pierced by the two fibers, a velocity can be computed since the distance between the two fiber extremes is known; $V_B = l_{12}/\Delta t_i \Delta t_i$ is the time interval between the fiber 1 and fiber 2. A method of associating the signals of the same bubble was developed since not every bubble is pierced by the both fibers. For this study almost 75% of the bubbles were associated [7], [8] and [9]. The average bubble velocity was hence obtained by,

$$\overline{V}_B = \frac{1}{N_b}\sum_i^{N_b} V_{bi}$$

and the velocity fluctuations obtained by:

$$\overline{V}_B'^2 = \frac{1}{N_b}\sum_{i=1}^{N_b}\left(\overline{V}_B - V_{bi}\right)^2$$

where Nb is total number of bubbles and Vbi is the ith bubble velocity (individual bubble velocity).

For every bubble detected on the two fibers the bubble chord length, depending on the position of the bubble on the probe was determined, $yi = Vbi \times \Delta t_i$. Where V_{bi} is the ith bubble velocity and Δt_i is the time interval of the bubble between fiber 1 and fiber 2. Again a histogram of bubble chords was constructed and the first moment of this distribution is the mean bubble chord as shown in Fig. 2.

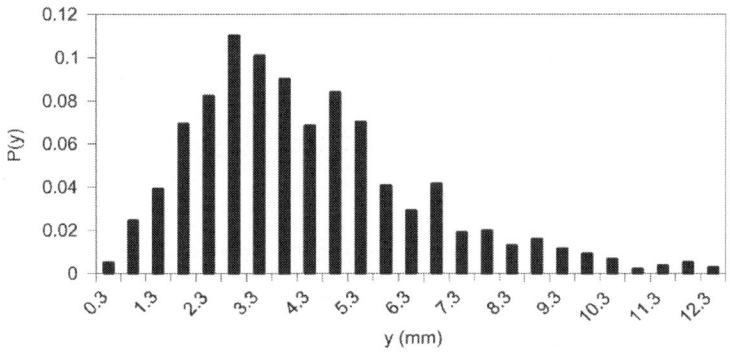

Figure 2: An example of histogram from the first probe.

By inverse transformation of the chords distribution through log-normal law we thus obtain the distribution of the bubble diameters [10], [11] and [12]. This is done by assuming that either the bubbles are spherical in shape or elliptical.

Hot Film Anemometry

This technique is based on the variation of temperature of the materials used to construct the film of the probe. According to King's law the convective heat transfer between the probe and the flow depends on the local Reynolds number and hence every probe can be calibrated using the Reynolds number and the Nusselt's number, $Nu = A + B.Re^{0.5}$. The current and hence the resistance of the probe is controlled by the Wheatstone bridge connected to an amplifier and a signal filter. In the work a constant temperature hot film anemometer was used. When a bubble passes on the film, there is a drop in convective heat transfer and this drop is reflected in the registered output signal. The signal treatment process is complex but gives access to local average liquid velocity, velocity fluctuations and the gas hold-up.

The hot film anemometer used in this work is a constant temperature anemometer DANTEC (Streamline) 55M01 equipped with a standard Wheatstone bridge 55M10. The equipment Streamline contains up to 6 probe compartments. The system has a driver which communicates permanently between the flow via the probe to the equipment. There is a temperature regulation sensor. The bridge is regulated until the signal is regular and stable. A single film probe (DANTEC 55R11).of 70 µm diameter and 1.25 mm length is used. An output signal as shown in the Fig. 3. The probe had to be calibrated before use. The data acquisition frequency is 2 kHz for 10–30 s giving rise to about 200 000 points per signal. The equipment has an acquisition card which analyses the raw signal in volts and converts it to velocity.

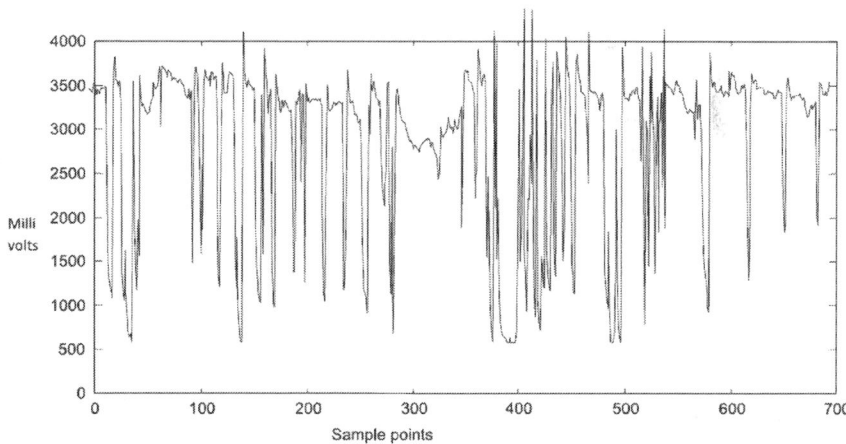

Figure 3: A raw signal of the hot film anemometer in two phase flow.

The signal from the anemometer consists of two sections a continuous part when the probe is in liquid and sharp drop when a bubble is pierced by the probe. The calibration setup was a vertical glass column of 50 mm in diameter and 1 m long connected to a tank of 80 liters and fed by a pump. Distilled water was used and the flow rate was measured by a flow meter. The tank's temperature was controlled by a thermostat at 30 °C±0.1 °C. A pitot tube which could be moved radially across the column connected to U-tube manometer inclined at 45° was used to measure the flow rate. The probe was placed at the same point with the Pitot tube and calibration was done. In fact the Dantec equipment could also automatically calibrate the probe from the velocities measured by the Pitot tube.

An analysis was carried out using Matlab program to cater for effects of the probe wetting by the liquid and to distinguish the bubble and the liquid phases. An example of the calibration curve is shown in Fig. 4.

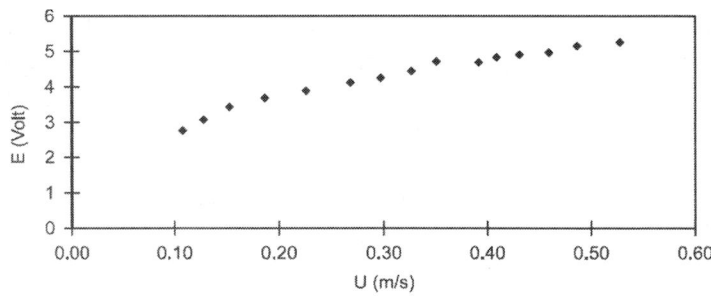

Figure 4: Calibration curve for the hot film anemometer.

Finally the liquid velocity was computed from the King's law

$$U(t) = \left[\frac{E(t)^2 - A}{B} \right]^{1/n} \text{[13]}.$$

EXPERIMENTAL SETUP

An external loop airlift was chosen to effect this work since it is commonly used in industries. It has the same advantages as the bubble columns but allows a better mixing between phases, hence higher material and energy transfer rates, they are easy to construct and the liquid circulation is self-induced hence no pumping is required. Its applications include the fermentation industry, water treatment, hydrogenation and oxidation, petrochemical and chlorination processes. Besides, in the processes where the by-products are gases such as CO_2, HCl or chlorine, these gases are easily removed at the top of the riser. An external loop airlift made of altu-glass with a height of 1.79 m, with the riser and down-comer of 94 mm and 50 mm diameter respectively was constructed as shown in Fig. 5. The horizontal distance between the two columns was 675 mm. The two columns were connected at the bottom by a tube of 50 mm in diameter. The elbows were rounded off smoothly (curvature radius of 125 mm) to reduce head losses. On the top the two columns were connected by a horizontal section open to the atmosphere, called the disengagement zone, allowing bubbles to escape before the liquid flows to the down-comer. The section was 747 mm long and 200 mm of height. The section tapered from 94 mm near the riser to 50 mm near the downcomer. The gas distributor was situated just above the riser, consisting of 8 parallel tubes of 6 cm in length having in total 56 holes of 0.6 mm diameter each. The distance between the holes was 11 mm. In the bottom a heater was installed to heat water and for the temperature regulation in the airlift. During all experiments the water temperature was maintained at 30.0±0.2 by a temperature sensor.

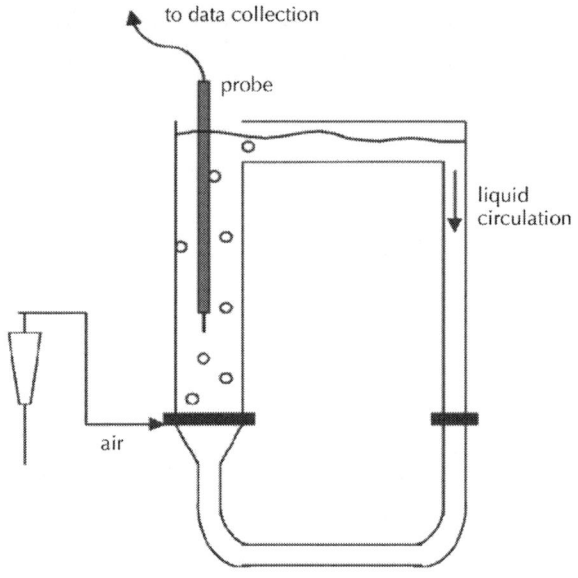

Figure 5: External loop airlift.

The liquid phase was distilled water; for all the experiments the water's unaerated height in the riser was 1.67 m corresponding to 16.5 liters. Compressed air was controlled by a rotameter calibrated earlier. The gas flow rates studied were 0.029, 0.047 and 0.069 m/s corresponding to homogenous, transition and turbulent regimes. The radial measurements were done at 5 axial heights in the riser at 0.10 m, 0.25 m, 0.5 m, 0.75 m and 1 m; corresponding to z/D=1.1,1.26, 5.3, 7.9 and 10.6 hence the flow conditions in the riser were fully explored.

RESULTS AND DISCUSSIONS

Local Gas Hold Up

The local gas hold up was measured by a bi-optical probe (ε_g). The axial and radial gas hold ups were explored. Plotted in the Figs. 6–7 are the radial profiles of gas hold-up at 4 dimensionless axial positions (z/D=1.1,5.3,7.9 and 10.5, with D=94mm) and for 3 gas superficial velocities ((U$_g$=2.9 cm/s, 4.7 cm/s and 6.9 cm/s).

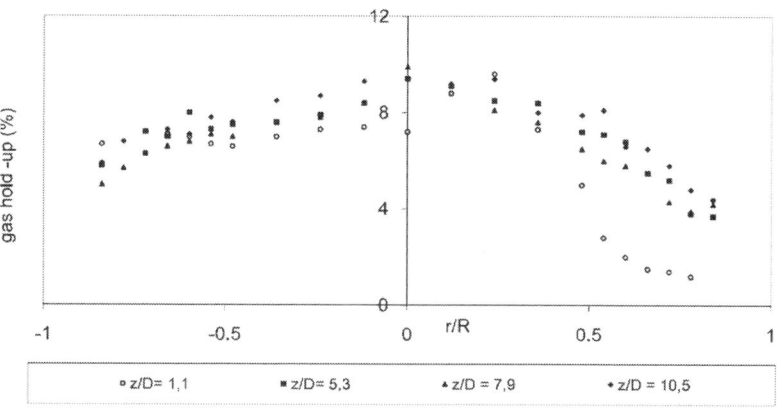

Figure 6: Gas hold-up profiles at U$_g$ = 2.9 cm/s.

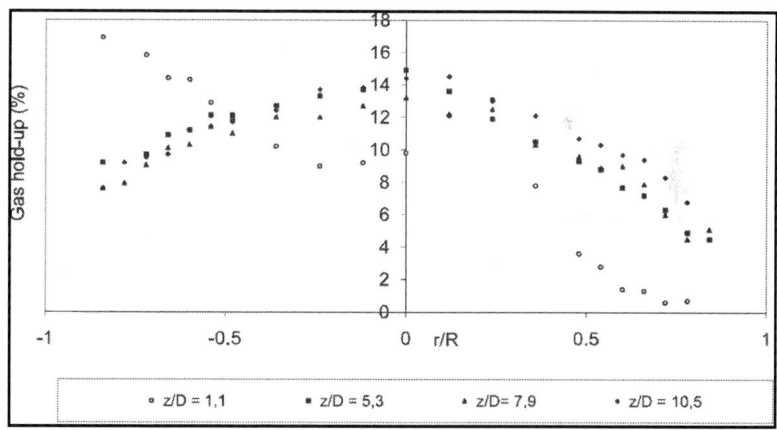

Figure 7: Gas hold-up profiles at t U$_g$ = 4.7 cm/s.

From the Fig. 6, Fig. 7 and Fig. 8 corresponding to z/D=1.1, one can notice an asymmetrical gas retention distribution at the base of the equipment. There is low gas hold-up on the right side (external part of the riser) and high gas hold-up on the left side (inner part of the riser). This observation is in agreement with works of [9], [14] and [15]. Since the gas distributor is uniform this is due to liquid flow around the elbow where bubbles are pushed by the liquid towards the inner side of the riser.

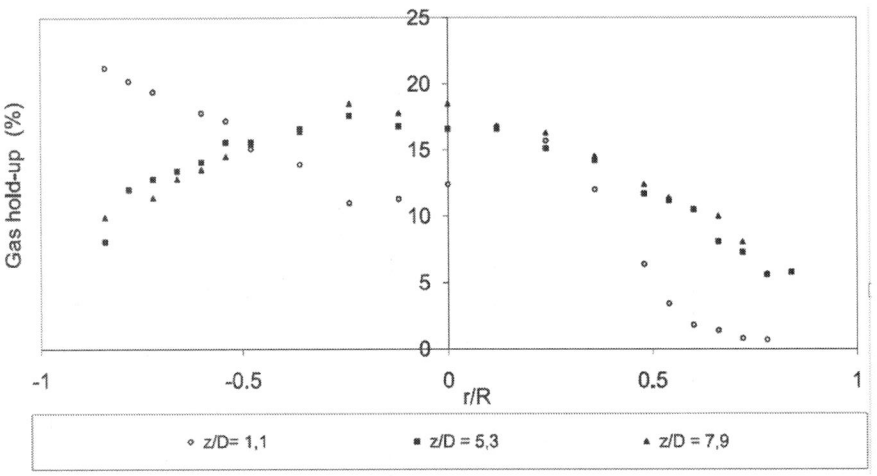

Figure 8: Gas hold-up profiles at U_g = 6.9 cm/s.

Above some distance in the riser the gas retention variation is moderate (about 12% between z/D=5.3and z/D=10.5 to U_g = 6.9 cm/s) since the bubble velocity controlled mainly by the liquid velocity should not vary much in the riser. Above 50 cm above the gas distributor (z/D=5.3) the radial gas retention profiles seem established and are parabolic in form. May be due to buoyancy effects the large bubbles migrate to the center of the column.

Fig. 9 shows the gas retention at 74.6 cm above the distributor for the three superficial gas velocities. It is noted that $_g$ increases with gas velocity, since the increasing gas velocity increases the gas volume in the riser. Between 2.9 cm/s and 6.9 cm/s of gas velocity, the gas hold-up increases by close to 60%. In fact the hold-up profiles are parabolic with higher values at the center. Similar results are reported by Vial et al. [16]and Wu et al. [17].

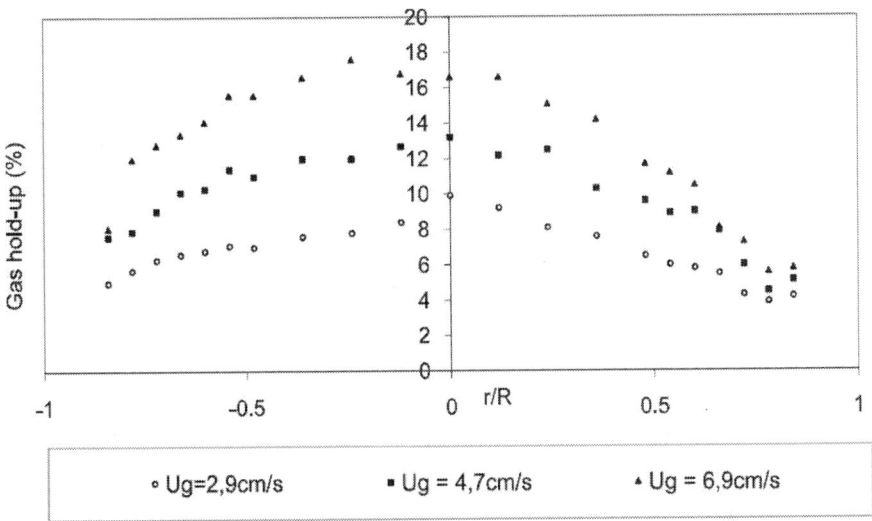

Figure 9: Radial gas retention profiles at z/D = 7.9(76 cm) above the distributor.

Bubble Sizes (Sauter's Mean Diameter)

The local Sauter's bubble diameter is deduced from the bubble chords' distribution and a statistical computation based on the hypothesis of the bubble shape (ellipsoidal), hence deducing the bubble diameter distribution. Log-normal bubble diameter distribution was found to be satisfactory. The assumption was found to hold so long as the number of bubbles pierced by the probe was more than 1000, which was the case for this experiment by both probes at high gas velocities. Fig. 10 shows an example of bubble diameter distribution obtained in the external loop airlift.

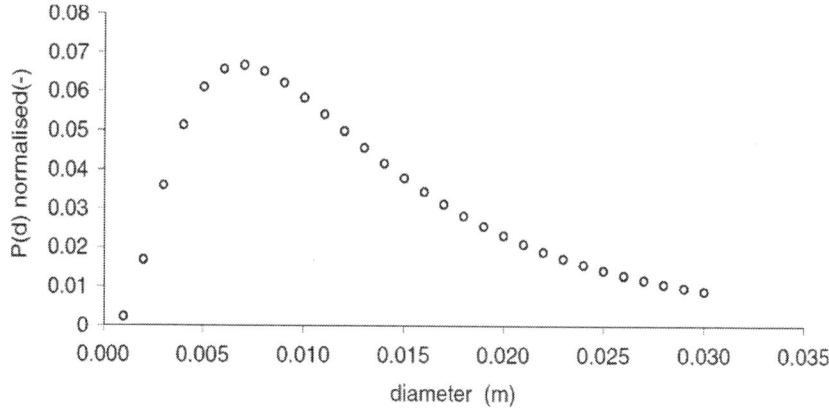

Figure 10: Normal bubble distribution obtained in the airlift reactor. Parameters of the normal log law: $\mu=0.004, \sigma=0.739$. Experimental conditions: Liquid phase = water; r/R=0; z/D=7.9; U_g = 6.9 cm/s.

Fig. 11 shows the radial Sauter's diameter $\left(d_{32} = 6\dfrac{V_p}{A_p} \right)$ profiles at 74.6 cm above the distributor for the 3 gas superficial velocities (2.9 cm/s, 4.7 cm/s and 6.9 cm/s). The bubbles' mean diameters increase with gas debit, with the largest bubbles (9 mm) found at the center. This conforms to the observation of the gas hold-up profiles where the highest levels are seen to be at the center of the riser due to buoyancy effects. The axial bubble diameter variations are shown on Fig. 12, Fig. 13 and Fig. 14 for the 3 gas superficial velocities. Again the parabolic forms suggest that large bubbles are found at the center of the riser. A strong concentration of large bubbles is found near the gas distributor (z/D=1.1) in Fig. 14; this conforms to the observation that the liquid current entrains bubbles with it which favors bubble coalescence as also reported by Utiger et al. [18].

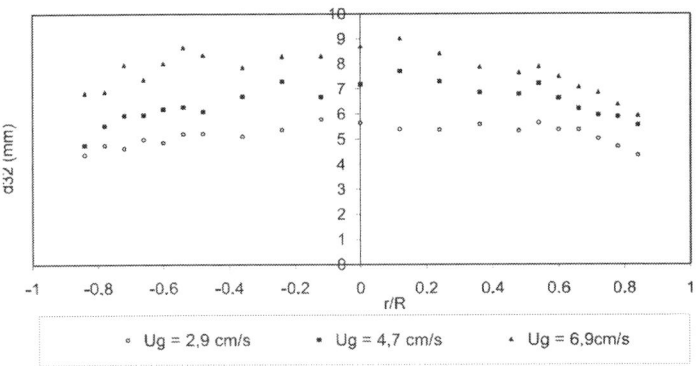

Figure 11: Radial Sauter's diameter profiles at z/D=7.9.

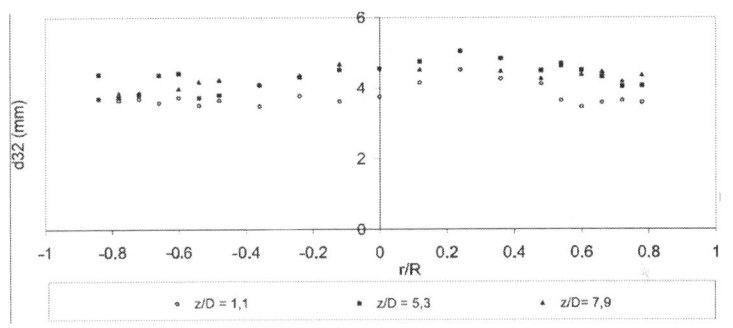

Figure 12: Radial Sauter's diameter profiles at t U_g = 2.9 cm/s.

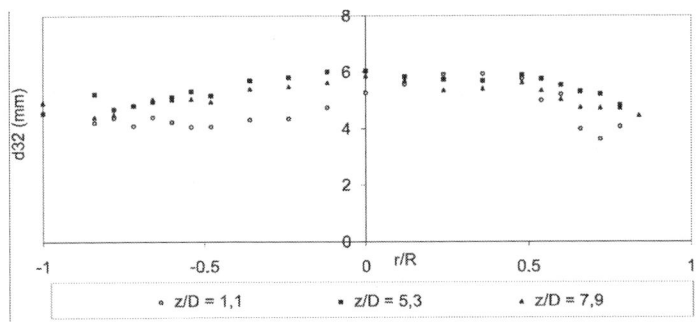

Figure 13: Radial Sauter's diameter profiles at U_g = 4.7 cm/s.

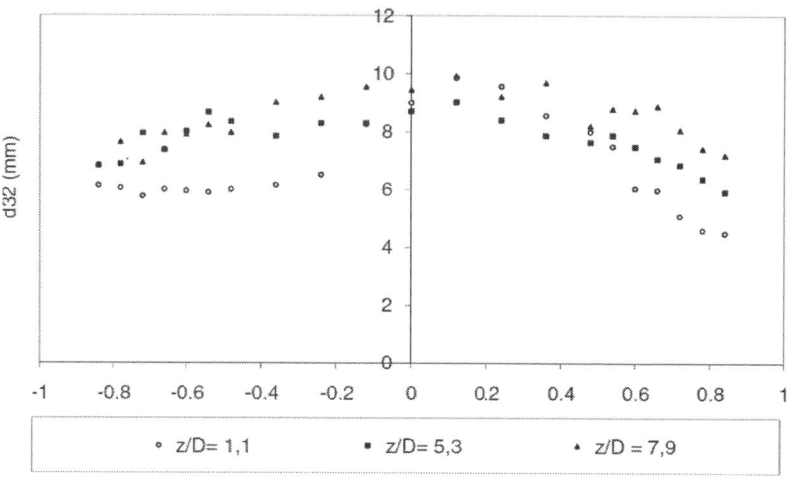

Figure 14: Radial Sauter's diameter profiles at U_g = 6.9 cm/s.

Bubble Velocities

The bubble velocities are determined by liquid velocities hence the higher the gas velocity, the higher the circulation liquid velocity and consequently the higher the bubble velocities. Fig. 15 shows the profiles atz/D=7.9, for the 3 superficial gas velocities. At this level the profiles seem established; the values increase with gas velocities. The parabolic form is due to liquid velocities but also the fact that large bubbles which rise faster due to buoyancy effects are found at the center. The variance of bubble velocity distribution at a given point (RMS fluctuations are shown in Fig. 16). Near the wall, where liquid velocity is low the fluctuations are weak getting stronger at the center where the liquid velocity variation is high.

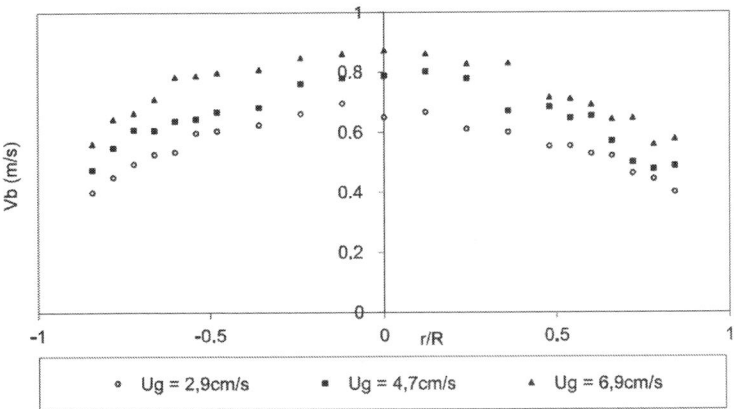

Figure 15: Radial profiles of bubble velocities at z/D=7.9.

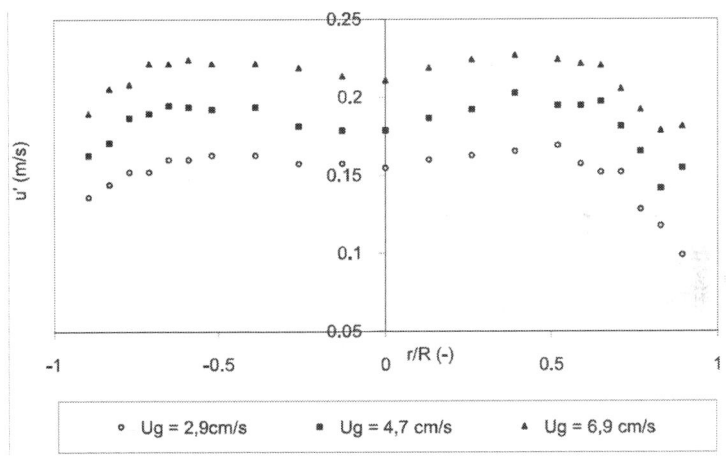

Figure 16: Radial profiles of variances of the distribution of bubble velocity at z/D=7.9.

Liquid Phase Characterization

The liquid axial velocities are measured in the riser by a hot-film anemometer as discussed in the experimental set up. The accuracy of the method is found in the works of Utiger [18].

The Local Mean Liquid Velocity

Fig. 17 corresponds to a gas superficial velocity of 6.9 cm/s.

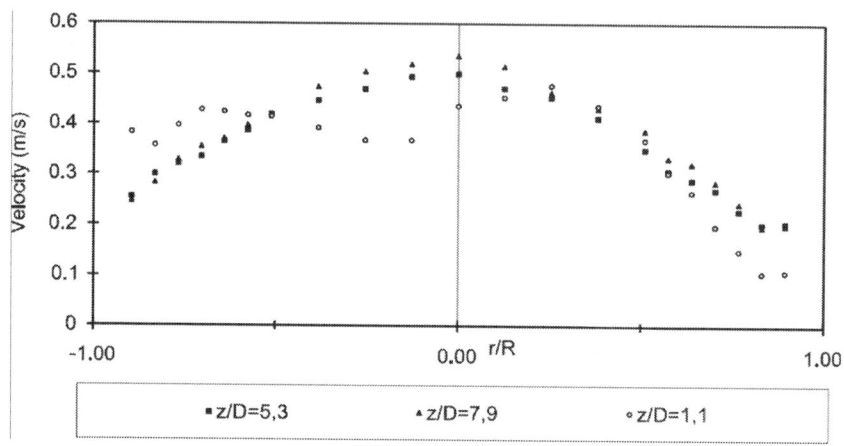

Figure 17: Radial liquid velocity profiles at U_g = 6.9 cm/s.

First of all at z/D=1.1 (just behind the distributor) the same asymmetry was noted on the liquid velocity distribution as was noted for the local gas hold up profiles. It is seen that most of the liquid current passes in the left half of the riser. This is of course due to the elbow in the riser. There could be some re-circulation of liquid in this section of the riser but this could not be verified as the hot film anemometer does not measure negative values of the velocity.

The liquid velocity profiles seem to be established from z/D=5.3 above the distributor see Fig. 18,Fig. 19 and Fig. 20. The parabolic form is again evident here as reported in a lot of work on local hydrodynamics in airlifts [16], [19] and [20]. The mean circulation velocity again increases with the gas superficial velocity: the difference in densities between the down comer and the riser is the driving force.

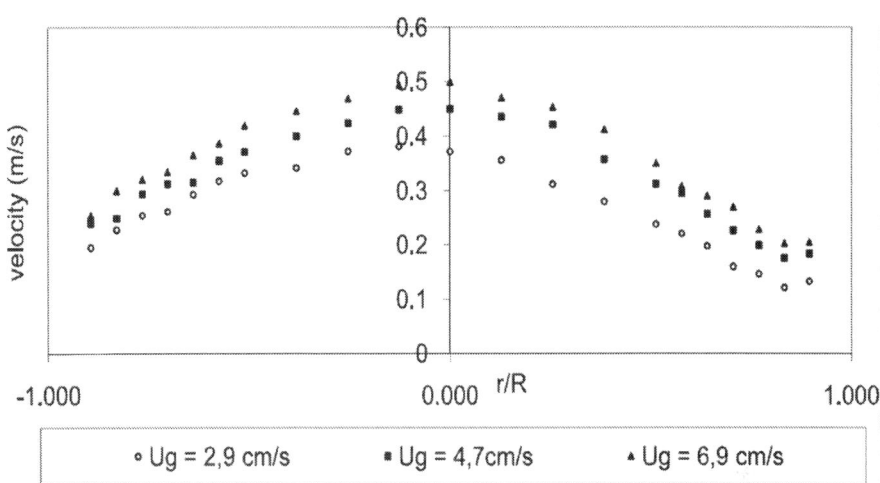

Figure 18: Radial liquid velocity profiles at z/D=5.3.

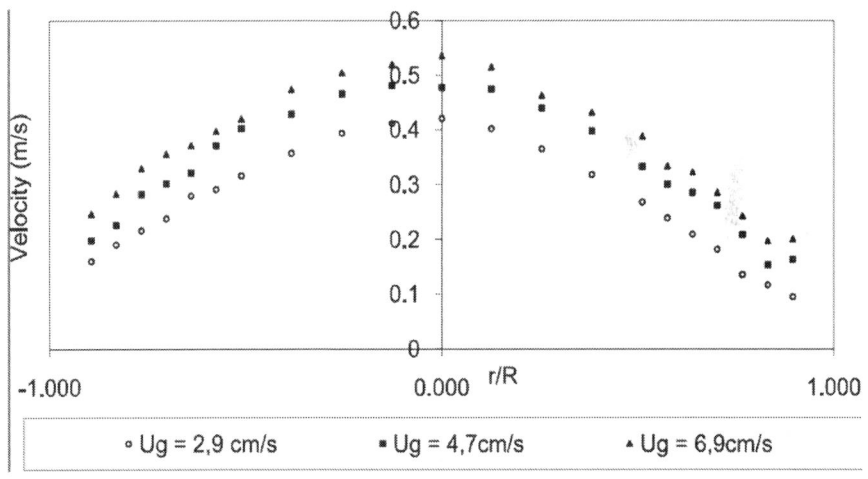

Figure 19: Radial liquid velocity profiles at z/D=7.9.

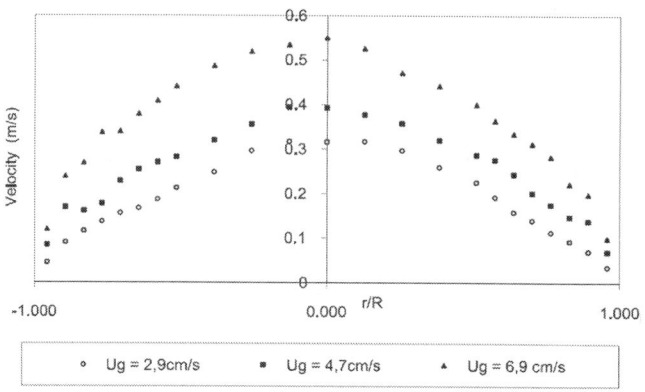

Figure 20: Radial liquid velocity profiles at z/D=10.5.

Liquid Velocity Fluctuations

The velocity fluctuations about the mean velocity were analyzed. Fig. 21 shows the radial profiles of the velocity fluctuations at liquid velocity of U_g = 6.9 cm/s The profiles seem established at a height half-way such as for other parameters. The fluctuations are weak at the center of the riser attaining a maximum value at mid-distance between the center and the wall to attain again weak values near the wall. At positionz/D=1.1 again the asymmetry is noted where the liquid current engendered by the elbow of the riser passes.

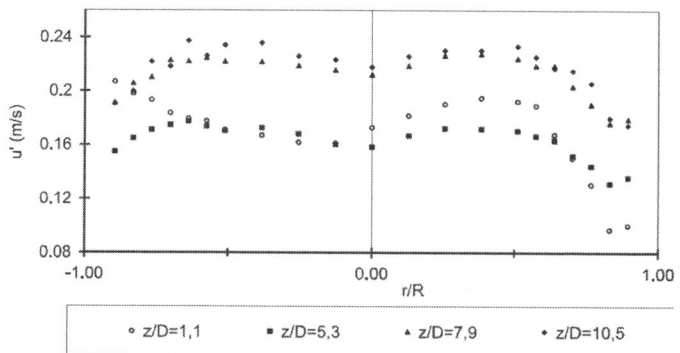

Figure 21: Liquid velocity fluctuations at U_g = 6.9 cm/s.

Fig. 22 provides values on the intensity of turbulence. Near the gas distributor the intensity is high again due to probable re-circulation of liquid. Above in the riser the profiles are again parabolic. The values are close to 30%, suggesting that the amplitude of the velocity fluctuations around the mean is significant.

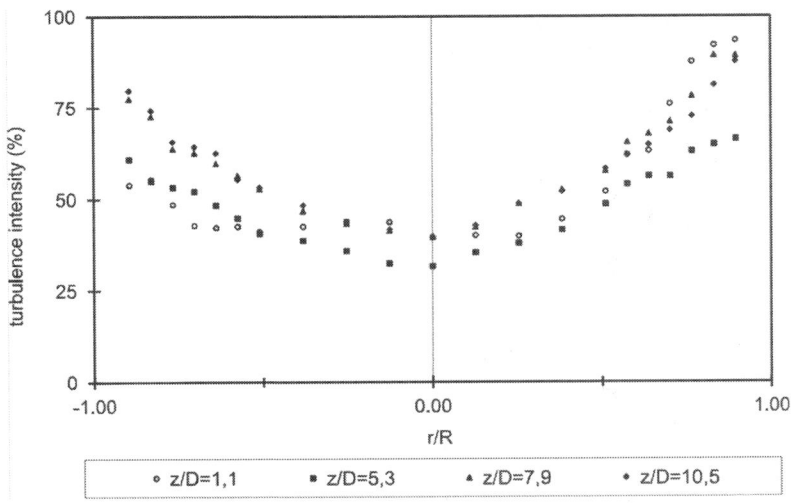

Figure 22: Profiles of turbulence intensity at U_g = 6.9 cm/s.

Slip Liquid Velocities

The bubble slip velocities (difference between the liquid and bubble velocities) are shown in Fig. 23. Before discussing these results it is important to note that this parameter depends on the accuracy of all other parameters (liquid velocity and bubble velocities). In the figure the slip velocity seems to increase with gas debit. This seems logical as an increase in gas velocity implies a high presence of bubbles which encourages bubble coalescence forming large bubbles with a higher velocity hence higher slip velocity.

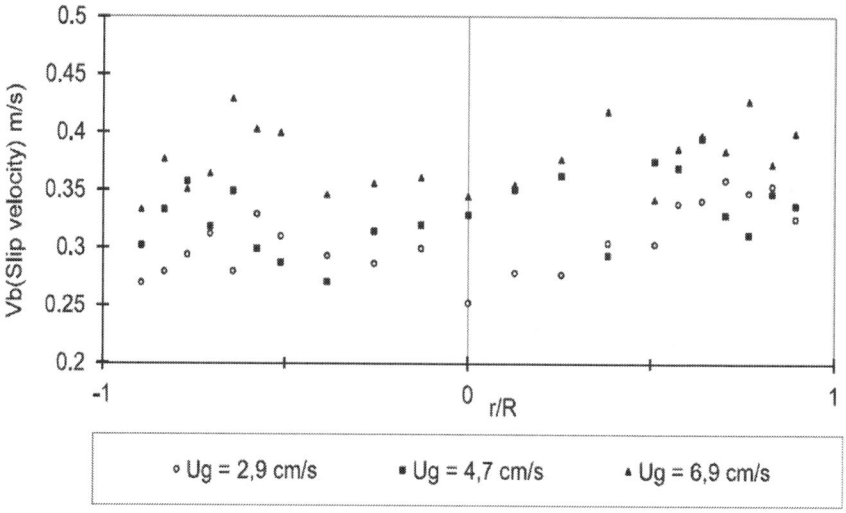

Figure 23: Radial profiles of gas slip velocity à z/D = 7.9 m/s.

CONCLUSIONS

In this study local hydrodynamics using intrusive methods in an external loop-airlift is presented. A bi-optical probe is used to study the dispersed phase in the riser while a hot film anemometer is used to characterize the continuous phase. In this reactor, we note a strong relationship between the two phases. Note that the liquid's circulation velocity is a function of gas velocity because of the density differences between the down-comer and the riser. The local parameters of both phases have been studied with reasonable accuracy. The experimental data generated can be used to validate simulation results (CFD). Besides it could be interesting to apply the same methods in other solutions having a different viscosity like sugars or tension-active solutions. The results agree with studies found in the literature [16], [19] and [20].

REFERENCES

1. Cartellier A, Achard JL. Local phase detection probes in fluid two-phase flows. Review of Scientific Instruments 1990;62:279–303.

2. Cartellier A, Barrau E. Monofiber optical probes for gas detection and gas velocity measurements: optimized sensing tip. International Journal of Multiphase Flow 1998;24:1295–315.

3. Boyer C, Duquenne AM, Wild G. Measuring techniques in gas–liquid and gas–liquid–solid reactors. Chemical Engineering Science 2002;57:3185–215.

4. Dias SG, Franca FA, Rosa ES. Statistical method to calculate local interfacial variables in two-phase bubbly flows using intrusive crossing probes. International Journal of Multiphase Flow 2000;26:1797–830.

5. Kamp AM, Colin C, Fabre J. Techniques de mesures par sonde optique double en l'ecoulement diphasique de bulles. Cert Onera. Colloque de Mecanique des Fluides Experimentale. Toulouse (France); 11–12 May 1995.

6. Chisti MY, Hallard B, Moo-Young M. Liquid circulation in airlift reactors. Chemical Engineering Science 1988;43(3):487–94.

7. Kiambi SL, Duquenne AM, Bascoul A, Delmas H. Measurements of local interfacial area: application of bi-optical fibre technique. Chemical Engineering Science 2001;56(21–22):6447–53.

8. Kiambi SL, Duquenne AM, Dupont JB, Colin C, Risso F, Delmas H. Measurements of bubble characteristics: comparison between double optical probe and imageing. The Canadian Journal of Chemical Engineering 2003;81:764–70.

9. Kataoka I, Ishii M, Serizawa A. Local formulation and measurements of interfacial area concentration in two-phase flow. International Journal of Multiphase Flow 1986;12(4):505–29.

10. Luo Hu-Ping, Al-Dahhan MuthannaH. Local characteristics of hydrodynamics in draft tube airlift bioreactor. Chemical Engineering Science 2008;63: 3057–68.

11. Kilonzo PM, Margaritis A, Bergougnou MA, Yu Y. Influence of the baffle clearance design on hydrodynamics of a two riser rectangular airlift reactor with inverse loop and expanded gas–liquid separator. Chemical Engineering Journal 2006;121:17–26.

12. Kilonzo PM, Margaritis A, Bergougnou MA. Hydrodynamic characteristics in an inverse internal-loop airlift-driven fibrous-bed bioreactor. Chemical Engineering Science 2010;65:692–707.

13. Zhang T,Wang J, Luo Z, Jin J. Multiphase flow characteristics of a novel internalloop airlift reactor. Chemical Engineering Journal 2005;109:115–22.

14. Klein J, Vicente AA, Teixeira J. Hydrodynamics of a three-phase airlift reactor with an enlarged separator—application to high cell density systems. The Canadian Journal of Chemical Engineering 2003;81:433–43.

15. Sánchez Mirón , Cerón García MC, García Camacho F, Molina Grima E, Chisti Y. Mixing in bubble column and airlift reactors. Chemical Engineering Research and Design 2004;82:1367–74.

16. Vial C, Poncin S, Wild W, Midoux V. Experimental and theoretical analysis of the hydrodynamics in the riser of an external loop airlift reactor. Chemical Engineering Science 2002;57(22–23):4745–62.

17. Wu X, Merchuk JC. Measurement of fluid flow in the downcomer of an internal loop airlift reactor using an optical trajectory-tracking system. Chemical Engineering Science 2003;58(8):1599–614.

18. Utiger M, StHuber F, Duquenne AM, Delmas H, Guy C. Local measurements for the study of external loop airlift hydrodynamics. The Canadian Journal of Chemical Engineering 1999;77(2):375–82.

19. Mudde RF, Van Den Akker HEA. 2D and 3D simulations of an internal airlift loop reactor on the basis of a two-fluid model. Chemical Engineering Science 2001;56(21–22):6351–8.

20. Zhonghuo D, Wang T, Zhang N, Zhanwen W. Gas holdup, bubble behavior and mass transfer in a 5 m high internal-loop airlift reactor with non-Newtonian fluid. Chemical Engineering Journal 2010;160:729–37.

7

An Integrative Image Measurement Technique for Dense Bubbly Flows with a Wide Size Distribution

Ashish Karn, Christopher Ellis, Roger Arndt, and
Jiarong Hong

Saint Anthony Falls Laboratory, 2 3rd Avenue SE, University of Minnesota, Minneapolis, MN 55414, USA

ABSTRACT

The measurements of bubble size distribution are ubiquitous in many industrial applications in chemical engineering. The conventional methods using image analysis to measure bubble size are limited in their robustness and applicability in highly turbulent bubbly flows. These flows usually impose significant challenges for image processing such as a wide range of bubble size distribution, spatial and temporal inhomogeneity of image background including in-focus and out-of-focus bubbles, as well as the excessive presence of bubble clusters. This article introduces a multi-level image analysis approach to detect a wide size range of bubbles and resolve bubble clusters from images obtained in a turbulent bubbly wake of a ventilated hydrofoil. The proposed approach was implemented to derive bubble size and air ventilation rate from the synthetic images and the experiments, respectively. The results show a great promise in its applicability for online monitoring of bubbly flows in a number of industrial applications.

INTRODUCTION

Bubbly flows occur frequently in natural systems and are also used for different applications in petroleum, energy-producing and chemical industries. Some of the common applications involve bubble columns which are used as reactors in a variety of chemical and biochemical processes, e.g. the Fischer–Tropsch process for hydrocarbon synthesis, hydrogenation of unsaturated oil, coal liquefaction, fermentation, waste water treatment etc. (Smith et al., 1996 and Lau et al., 2013). Bubbly flows are also ubiquitously found in flotation cells (Sadr-Kazemi and Cilliers, 1997), aeration studies (Roesler and Lefebvre, 1989) and spargers (Geary and Rice, 1991) etc. In many of such processes, the accurate prediction of pressure drop and wall heat transfer is necessary, both of which are strongly dependent upon the concentration, spatial distribution and morphology of the bubbles (Kamp et al., 2001). Similarly, in many liquid–gas systems, gases are dispersed in liquids to obtain large interfacial area available for chemical reactions, heat and mass transfer processes. The rate of such processes is characterized by bubble surface area flux which is closely associated with the bubble size distribution (Junker, 2006).

Different techniques have been employed to measure bubble size distributions. Broadly, it can be divided into two categories –

intrusive and non-intrusive techniques. Both these methods have been extensively reported in the literature – some of the intrusive methods employ capillary suction probes (Laakkonen et al., 2005), conductivity probes (Liu and Bankoff, 1993), optical fiber probes (Saberi et al., 1995) and wire-mesh sensors (Prasser, 2008), etc. The non-intrusive methods include interferometric particle imaging (Glover et al., 1995), laser Doppler velocimetry (Mudde et al., 1998), extinction and scattering activity measurement (Zaidi, 1998), phase Doppler anemometry (Laakkonen et al., 2005) and other particle-imaging techniques (Tayali and Bates, 1990, Adrian, 1991 and Grant, 1997), etc. In general, non-intrusive methods are preferred over intrusive methods which disturb the local flow fields because of the placement of the probes.

Digital image analysis offers many advantages in terms of flexibility, relative insensitivity to the optical properties of the dispersed phase, easier optics alignment as compared to laser-diffraction methods, as well as the capability of providing the velocity and size information of the dispersed phase simultaneously. Thus, it is very convenient and time efficient for online monitoring and analysis of a large number of images. However, to implement this technique for real-time analysis of the bubbly flow images from different engineering applications pose multifarious challenges. These challenges include, for instance, the computational speed for real-time image processing, the ability to cope with the poor quality of images caused by varying intensity characteristics of the background and out-of-focus objects, and the robustness of the technique especially in its capability to resolve overlapped clusters in the high void-fraction flows.

The optical image analysis have been used recently for quantifying the bubble size distribution (e.g.Honkanen et al., 2010, Ferreira et al., 2012, do Amaral et al., 2013, Kracht et al., 2013 and Lau et al., 2013). Generally, due to the excessive coalescence and break-up of bubbles, most of the proposed techniques for bubble image processing produce considerable errors when applied to flows with high superficial gas and liquid velocities. These errors are closely related to the challenge of extracting accurate bubble information from large clusters due to the coalescence of bubbles.

A brief review of these techniques is presented inSection 3.1. Overall, these techniques are still limited in their robustness to resolve large bubble clusters particularly under highly unsteady flows with large void fractions of bubbles. Another limitation of the reported techniques is related to their ability to deal with a wide range of bubble size distribution. In addition, algorithms with significant improvements in computational speed are needed for fast processing of a large number of images and online monitoring of bubble concentration and distribution.

Thus, in the present study, we introduce an integrative image measurement technique to analyze high void fraction bubbly flows with a wide dynamic size range of bubble size. The development of this technique is driven by our recent study on the bubbly wake flows of aerated hydrofoils. This research is focused on developing a test-bed through conducting physical water-tunnel experiments to quantify the dissolved oxygen transfer across bubbles under various flow conditions. The experiments result in a large quantity of varying quality of bubble images with significant clustering due to highly unsteady and complex flows and coalescence of bubbles. These images make it unfeasible to implement prior measurement techniques to achieve fast and accurate image analysis.

This paper is structured as follows: Section 2 provides the details for the experimental facility, setup and the optical approach used to capture digital images. The proposed image analysis algorithm is described inSection 3. Subsequently in Section 4, we present validation of our image analysis technique through both simulation and experimental approaches, which is followed by a final conclusion in Section 5.

DESCRIPTION OF EXPERIMENTAL SETUP

The experiments were conducted in the high-speed water tunnel at Saint Anthony Falls Laboratory (SAFL) of the University of Minnesota.

The tunnel has a horizontal test section of 1 m (Length)×0.19 m (Width)×0.19 m (Height) with three sides having plexiglass wall for optical access. The tunnel is designed for cavitation and air ventilation studies and is capable of operating with velocity in excess of 20 m/s. A special design feature of the tunnel provides for fast removal of large quantities of air bubbles generated during cavitation and ventilation experiments, allowing us to conduct bubbly flow experiments for extended periods of time with little effect on test section conditions.

During the experiments, a NACA0015 hydrofoil was installed in the test section with angle of attack (α) of 0°, 4° and 8°. The hydrofoil was 190 mm in span and 81 mm in chord. As shown in Fig. 1, a narrow spanwise slot allows air to be injected into the flow over the hydrofoil. The full width of the injection slot is used for measurements of oxygen uptake. This results in a dense spanwise bubbly wake. However, in order to make bubble measurements, ventilation was limited to a narrow 9.6 mm slot at the center of the span. This configuration ensured that bubbles remain mostly within a narrow distance away from the center. Considerable thought was given to obtain a reasonable representative sample of the bubble population that exists when the full span is ventilated. Under this scheme, 45 different experiments were conducted at different water speeds, ventilation gas flow, angles of attack of hydrofoil and the bubbly wake images were obtained at three different axial locations, i.e. 109, 243 and 377 mm from the hydrofoil center. 90,000 bubble images were captured using the SIV technique.

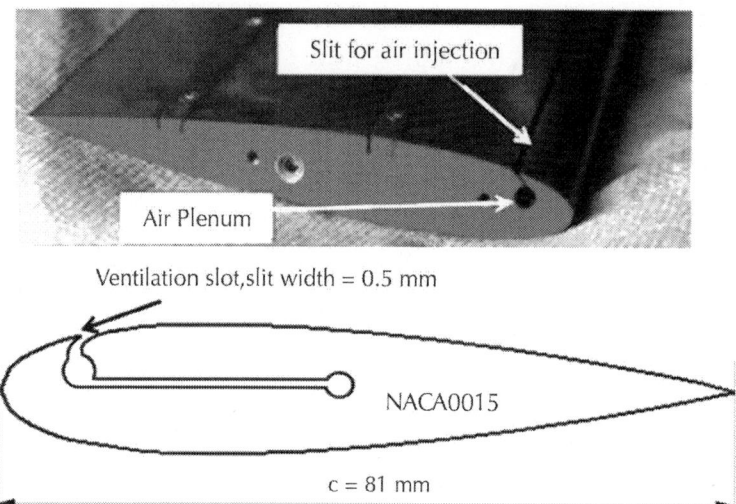

Figure 1: Details of the ventilated foil.

Shadow Image Velocimetry technique (SIV) is most suited for analysis of bubbly flows. It employs direct in-line volume illumination using low power sources such as LED and an optical setup to produce a narrow depth-of-field for 2D plane imaging (Goss et al., 2007 and Bröder and Sommerfeld, 2007). Fig. 2 shows a schematic of the experimental setup. A custom-made pulsed LED light source from Innovative Scientific Solutions Inc. was used to illuminate the flow. The pulsed LED array has flash rates up to 10 kHz with a 5 µs pulse width and rise and fall times ~200 ns. To ensure uniform back-lighting in the images, a light shaping diffuser is placed between the light source and the flow, which eliminates noise generated by non-uniform lighting. A 1 K×1 K pixel Photron APX-RS camera (capable of 3000 frames/s at full resolution) with a 60 mm lens was used to obtain images.

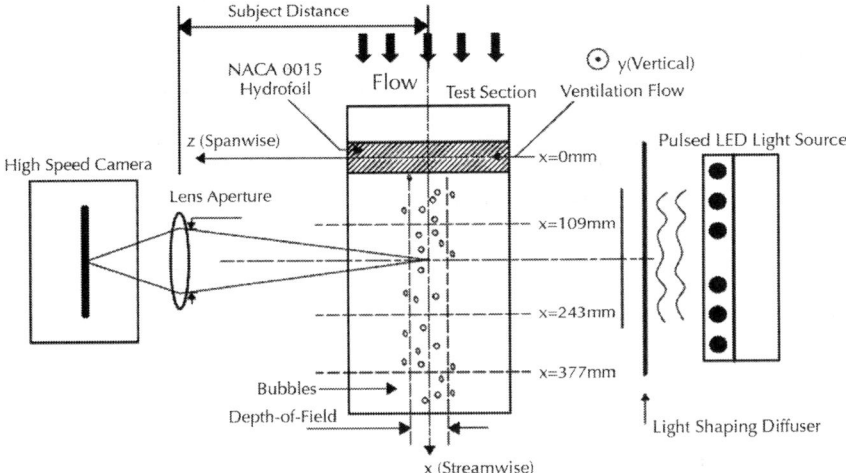

Figure 2: Schematic of the experimental setup for shadow imaging of bubbly flows in an aerated hydrofoil wake.

In the SIV technique, two LED light pulses separated by a short time are synchronized with camera exposure in order to obtain two consecutive images (or, double frames). The first pulse is fired at the end of the first exposure and the next at the beginning of the following exposure. In our experiments, the time duration between two pulses was kept 100–230 µs depending on free stream velocity. Using the obtained image pairs, the instantaneous velocity field of the bubbles was obtained using commercially available Particle Image Velocimetry (PIV) software (DaVis 7.2 from LaVision). The image pairs were captured at a frame rate of 25 image-pairs per second and the exposure time for individual images was kept 15 µs to prevent any blurring in the images. A data set consisted of 1000 image-pairs taken over 40 s. The field of view of the captured images was approximately 60 mm×60 mm. The imaging system was calibrated prior to the beginning of the test program using a 2.5 mm×2.5 mm grid located at the water filled test section centerline. The calibrated images produced by the test program and used in the analysis described here had a constant (square) pixel dimension of 0.059 mm. As mentioned, this pixel dimension was determined by a grid placed at the test section centerline. The bubble plume exited the foil with a spanwise depth equal to the non-masked slot length of 9.6 mm and spread to at most twice this depth at the downstream measurement location. This uncertainty of spanwise bubble location translated to a length scale (and calculated velocity) uncertainty of 1.6% near the foil and 3.2% at the downstream measurement location. The image depth of field was determined to be approximately 15 mm. The bubble sizes were in the range of 2–68 pixels, which corresponded to a bubble radius of 0.06–2 mm.

PROPOSED METHODS

Overview of the Image-processing Task

Fig. 3 presents a sample of the original bubble images obtained from our measurements to illustrate a number of challenges involved in the analysis of our images. These challenges include (1) the spatial and temporal non-uniformities of the image background; (2) a wide range of bubble size distribution; (3) the co-existence of in-focus and out of focus bubbles; and (4) the large bubble clusters.

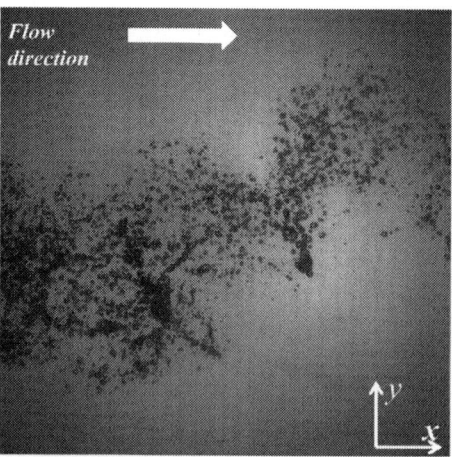

Figure 3: A sample of original bubbly wake image from our experiments.

This spatial non-uniformity, shown as a relatively brighter central disk in Fig. 3, is generated by the non-uniform spatial profile of the LED illumination. The temporal non-uniformity in the background is contributed by the fluctuations of LED illumination, and more significantly, by the fluctuating void fraction of bubbles associated with highly unsteady turbulent flows in the field of view. In our experiments, the high ventilation rate and high Reynolds number turbulent flow enhance the bubble break-up and coalescence, which leads to a large dynamic range of bubbles including micro-bubbles to large clusters of several bubbles. In addition, the complexity and three dimensionality of turbulent flows result in a substantial amount of out-of-focus bubbles.

In our case, the conventional methods reported in the literature to deal with these challenges proved to be inadequate. Usually, a background subtraction can be employed to compensate for the temporal non-uniformity. However, since the background illumination intensity varies both with location and time, the background subtraction can lead to substantial loss of bubble information.

Similarly, the spatial non-uniformity caused by the uneven illumination poses a problem for image filtering using either a global, local thresholding or a combination of both. Specifically, for image-filtering using a local thresholding based on block-processing with a suitable filter function, different filter functions have been used in the literature for the determination of threshold value (e.g. Otsu, 1979, Sahoo et al., 1997 and Sahoo and Arora, 2004). In the current research, these different filter functions were tried on a variety of images, but were found to be unsatisfactory. It was observed that huge variations in the local void fraction in different images rules out the possibility of a single threshold value for every image or even for small local regions in the image.

These drastic differences in the local void fraction can be attributed to the presence of dense bubble clusters. Usually, bubble clusters are very difficult to deal with and require specific approaches for analysis. Sometimes, these clusters are ignored in the analysis by constraint conditions such as sphericity (Bailey et al.,

2005) or concavity index (Mena et al., 2005). From a classical image analysis point of view, it is assumed that bubble clusters occur as a non-selective process and ignoring these clusters would not bias the measurement. However, recently Kracht et al. (2013) showed through a stochastic approach that even if the bubble clusters are non-selective, large bubbles are more likely to be present as clusters. Thus, ignoring these clusters in the measurement would bias the estimations.

To consider the size of such bubble clusters, some authors have proposed to approximate the overlapping bubbles through an object recognition approach which fits an ellipsoidal shape to the object areas (Pla, 1996 and Honkanen et al., 2005). Some other reports have focused upon segmenting these clusters by implementing watershed algorithm (Bonifazi et al., 1999, Lin et al., 2008, Zhou et al., 2010, Zhang et al., 2011 and Lau et al., 2013). However, a substantial over-segmentation was observed in many cases upon using watershed transform. Such over-segmentation is usually avoided by suppressing the shallow minima using H-minima transform (Eddins et al., 2004).

However, the different variability in the image makes the exact definition of "shallow" minima difficult. In addition, the application of watershed transform requires a pre-processing step to overcome the problems of over-segmentation, which can eliminate small size bubbles in a bubbly flow with a large dynamic range of bubble size.

The small size bubbles, because of their large surface area to volume ratio are extremely crucial in our experiments because of their significant contribution to the oxygen transfer as compared to the large size bubbles. Also, there are a large number of small bubbles in the flow field and thus these cannot be ignored.

Finally, it was concluded that to effectively extract the bubble information over a wide size range, to overcome the different non-uniformities in the images and to deal with both in-focus and out-of-focus bubbles, a robust and integrative algorithm is needed.

Image Processing Approach

Summary of the Algorithm

The proposed approach assumes bubbles are generic ellipsoids and thus the bubble properties such as centroid location, size and shape can be extracted from the projected area of a bubble. Fig. 4 shows the basic outline of the image analysis procedure. As shown in the figure, the image analysis consists of three major steps: (1) Image binarization: the original grayscale images are converted into binary images using extended H-Minima transform with a suitable parameter (details presented in Section 3.2.2); (2) Bubble categorization: all the bubble regions are labeled and characterized by their position and a series of metrics including area, centroid, major and minor radius and circularity factor. Based on the area of each region, the bubbles are divided into tiny spherical bubbles, intermediate-size bubbles and large bubbles/clusters. (3) Bubble information extraction: Owing to the different sizes and characteristics of bubbles in the images, a single universal approach cannot be used to extract the size information of all the bubbles. Thus, a multilevel segmentation approach is suggested to extract the maximum possible information from the images. A customized approach referred to as 'Cluster Processing' (see Section 3.2.3 for details) was developed using advanced morphological operations and a watershed transform to extract individual bubbles from bubble clusters.

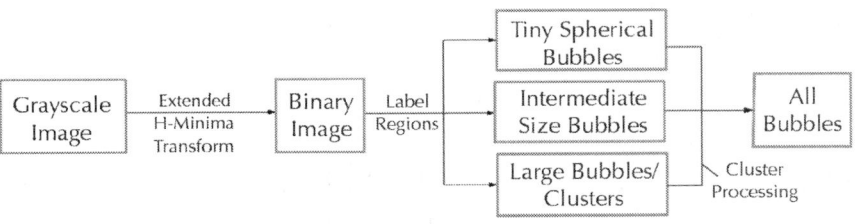

Figure 4: Basic outline of the image-processing technique.

Extended H-minima Transform

H-minima transform of a grayscale image suppresses all minima in the image whose depth is less than a particular value (Soille, 2003). Here, the depth of a minimum refers to the difference in grayscale intensity value of background and the bubble. A unique feature of this technique is that thresholds are applied to the depths of minima and not the grayscale intensity value in itself. Extended H-minima transform is the regional minima of the H-minima transform, where a regional minima is defined as connected components of pixels (8-connected) with a constant intensity value whose external boundary pixels all have a higher value. Thus, the extended H-minima transform of a grayscale image produces a binary image with '0' representing the background and '1' representing the bubble region. It is important to note that the extended H-Minima transform of an in-focus bubble produces a hole in the bubble region owing to the central bright portion in the bubble. Since extended H-minima transform computes the regional minima and does not require a fixed pre-defined global threshold for the grayscale intensity value for conversion of grayscale image to binary image, it is more robust to the variability in the image caused by fluctuations in background illumination intensity and also because of uneven illumination in the background. Also, the binarization process is not extremely sensitive to the depth-of-minima threshold.

Cluster-processing

Fig. 5 illustrates our approach to extract individual bubble information from bubble clusters. A bubble cluster might contain both in-focus and out-of-focus portions. As shown in Fig. 5, the in-focus and out-of-focus portions of a bubble can be distinguished by a central dark hole, which is generated from the central bright portion in the original grayscale images of in-focus bubbles (see Fig. 3). However, for the out-of-focus bubbles, only the bubble boundaries can be traced. It is important to note that the narrow ventilation slot in our experiments ensures that bubbles stay near the depth-of-field of our measurements as discussed in Section 2. Based on the characteristic features of in-focus bubbles, a bubble cluster is divided into the in-focus and the out-of-focus portions and different techniques are employed to segment each portion into individual bubbles. Specifically, first, a set of morphological operations are applied to a bubble cluster (shown in Fig. 6) containing both in-focus and out-of-focus portions. These morphological operations resolve only the in-focus bubbles and the out-of-focus bubbles are eliminated in the process. Subsequently, the out-of-focus portion is obtained by subtracting in-focus separated individual bubbles from the original bubble cluster and resolved separately into individual bubbles.

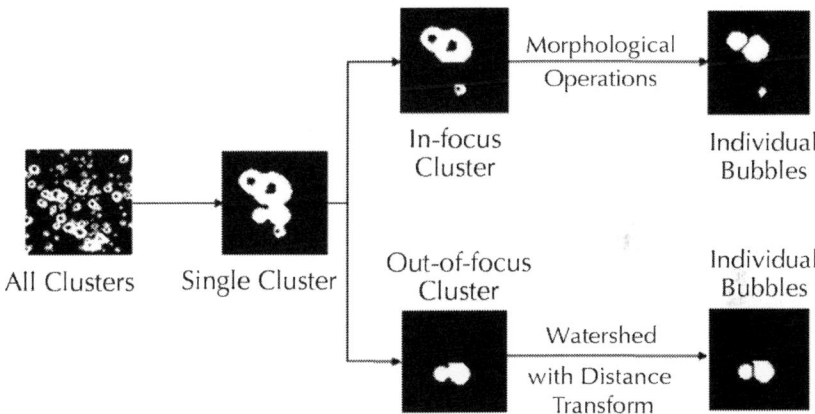

Figure 5: A detailed outline of steps in the 'cluster-processing' technique for a hybrid cluster.

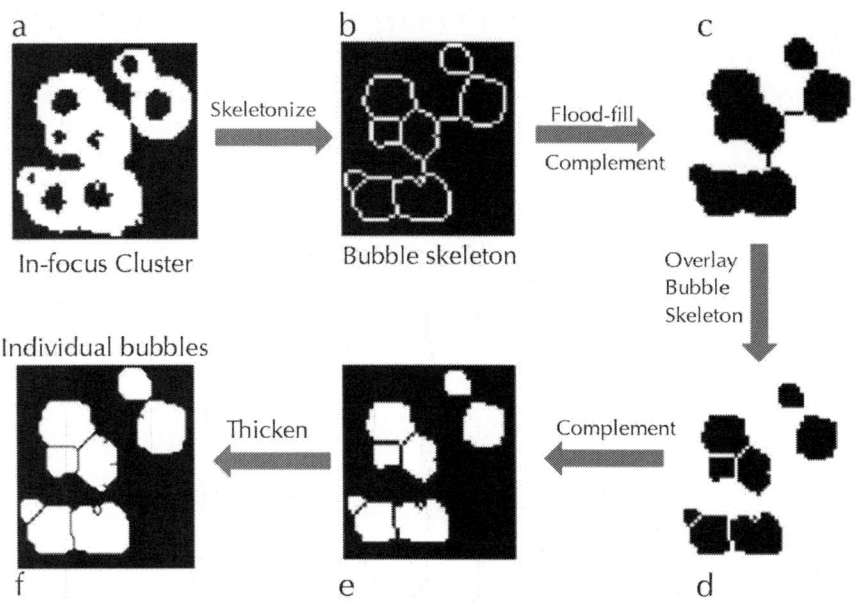

Figure 6: Cluster-processing technique for the in-focus bubble-clusters.

Fig. 6 describes the set of morphological operations that were employed to resolve an in-focus cluster into separate bubbles. First, a single cluster is selected with a bounding box around its centroid. Portions of the other contiguous clusters that fall into this bounding box are eliminated based on an area threshold (Fig. 6a). The skeletonization of the chosen cluster reduces the foreground regions in the cluster to a skeletal remnant that preserves the shape and connectivity of the original region (Fig. 6b). This is followed by a morphological 'shrink' operation, which is iteratively performed on the cluster till a point when there is no further change in the skeleton. A shrink operation combined in succession with an area-opening operation removes all the foreground pixels and retains the pixels around the hole. Next, a flood-filling operation was carried out and a complement was taken to obtain a cluster where the boundaries of bubbles are not visible (Fig. 6c). The bubble skeleton is then superimposed to obtain the separated bubbles (Fig. 6d). Subsequently, a complement and thicken operation was done to offset the effect of iterative shrink operation and finally obtain the separated individual bubbles (Figs. 6e and f). The structuring element size for the 'thicken' operation was chosen in a way so as to ensure that the sum of all separated individual bubble areas equals the total filled area of cluster. Note that, due to inherent nature of the proposed technique there may be an error of a pixel in resolving the boundary between two different bubbles. Thus, to obtain the actual size of the bubbles, a boundary correction has to be added after the measurements of the bubble size from Fig. 6f.

The proposed technique essentially depends on the intensity gradients in the bubbles and is limited in resolving the out-of-focus portions of a cluster where intensity gradients are not clearly visible. For this reason, the out-of-focus cluster is resolved into separate bubbles by employing a different technique: watershed segmentation with distance transforms (Chen et al., 2004). The watershed transform is a region-based segmentation technique originally proposed by Meyer (1994), which is based on an analogy with the flooding of a landscape or topographic relief with water (Lin et al., 2008, Zhou et al., 2010 and Zhang et al., 2011). This technique requires pre-processing the images by edge detectors and smoothing filters and can otherwise lead to over-segmentation (Bonifazi et al., 1999). Considering the different sources of variability in the image, it is difficult to use watershed segmentation to resolve all the different sized in-focus bubbles accurately without sacrificing some small size bubble information. Thus, in the current research, we leverage separate approaches for in-focus and out-of-focus bubbles to extract maximum information from our bubble images.

Bubble Size and Shape Measurements

The bubble information including the centroid location, size and shape is extracted based on the projected areas of each bubble in a bubble image. The bubbles are categorized as being circular or elliptical in cross-section and finally bubble size is measured and converted from pixel to metric values using the magnification of the image.

First, all the bubble regions were grouped into three categories based on size: tiny bubbles, intermediate size bubbles and large bubbles/clusters. Based on the observations in our experiments, it is clear that very small bubbles are almost always spherical, i.e. have a circular projected area in our images. The area threshold for such tiny bubbles was set to 10 pixels. For these circular bubbles, the radius was computed directly from the area measurements (A) as $d = \sqrt{4A/\pi}$. The intermediate size individual bubbles (10–30 pixels) and the bubbles resolved after 'cluster-processing' (>30 pixels) could have a circular or elliptical projected area. Thus, the bubble shape needs to be determined before the bubble sizes are calculated. The bubble shape is determined by the calculation of Heywood Circularity Factor, defined as $HCF = P/\sqrt{4\pi A}$, where P is the perimeter and A is the area of the region. The Heywood circularity factor is defined as the ratio of bubble perimeter to the perimeter of a circle of the same area. For each bubble, a Heywood circularity factor was calculated. Note that spherical bubbles have a HCF of unity whereas a square region has a HCF of 1.128. Again, a conservative range of HCF for circular bubbles was selected to be 0.9–1.15. For bubble regions lying within this range of HCF, the radius was determined using the assumption of a spherical bubble. For other bubbles lying outside this range of circularity factor, an ellipse fitting technique proposed by Haralick and Shapiro, (1992) was used to determine minor and major axis lengths and the eccentricity of the elliptical projected area.

RESULTS AND VALIDATION

The proposed image processing technique was validated using both synthetic bubble images and air ventilation rate measured from the experiment.

Validation through Synthetic Bubble Images

To quantify the precision of bubble size and shape capture, we first implement our approach to synthetic image of a single bubble. Then, polydispersed bubble images that simulate the bubble distribution in our experiments are generated to test the effectiveness of our approach in extracting bubble information from large bubble clusters.

Synthetic Bubble Generation Procedure

The synthetic bubble images are generated using an ensemble of randomly distributed ellipses. The size, eccentricity and orientation of the elliptic objects are also randomly selected within the range of bubble characteristics measured from the experiments. As illustrated in Fig. 7, an intensity profile, approximated using the quadratic function below, is superimposed on each ellipse to reproduce the intensity profile of bubbles from the experiments. The intensity at any radial position, r is defined by the quadratic:

$$I = ar^2 + b$$

where

$$a = \frac{(I_m - I_i)}{R^2} \; ; b = I_i \qquad\qquad if \;\; |r| < R$$

$$a = \frac{(I_b - I_m)}{(R_2^2 - R^2)} \; ; b = I_m - \frac{(I_b - I_m)R^2}{(R_2^2 - R^2)} \qquad if \; R_2 > |r| > R$$

I_m, I_i and I_b correspond to different fixed pixel intensity values in a bubble image as shown in Fig. 7. *Im* refers to the minimum intensity in the bubble cross-section, *Ii* is the intensity of the central bright portion of the bubble and *Ib* refers to the background intensity of the bubble images. The bubble radius is given by $R_2 = R+w$, where w is bubble width. The bubble centroid is chosen as the origin of the co-ordinate system. Thus, a generic elliptical synthetic bubble image was synthesized with the prescribed intensity profile that matches that of an actual bubble.

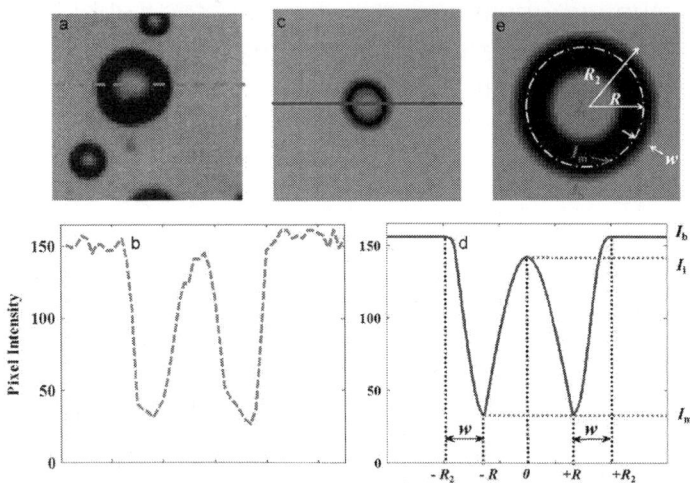

Figure 7: (a) An actual bubble image and (b) its corresponding intensity profile is used to replicate (c) a synthetic bubble image and (d) its intensity profile. (e) A synthetic bubble image showing different parameters involved in bubble generation.

Single Bubble Measurement

To validate the proposed image analysis algorithm, we employ a methodology similar to that reported byLau et al. (2013). First, we apply the image processing algorithm on single synthetic bubbles for different sizes and shapes (i.e. circular and elliptical). An elliptical shaped bubble is drawn with a specific radius in pixels and its centroid is placed in the middle of a pixel (0.5 pixel, 0.5 pixel). Subsequently, 50 new bubbles with same size and shape are generated with its centroid translated with steps of 0.01 pixel in both x and ydirection. For each translation, the image analysis algorithm calculates the bubble equivalent radius and the percentage error in measurement of bubble area-equivalent radius is calculated. The bubble area-equivalent radius is the geometric mean of the minor and major radius of a bubble. The average size and range is presented as error bar in Fig. 8. The same process is repeated for bubbles of different radius ranging from 5 to 105 pixels and for both elliptical (eccentricity=0.6) and circular bubble shapes. Fig. 8 presents a comparison of the error in radius measurement for the circular bubbles, elliptical bubbles and the maximum possible uncertainty in the measurement estimated by Lau et al. (2013).

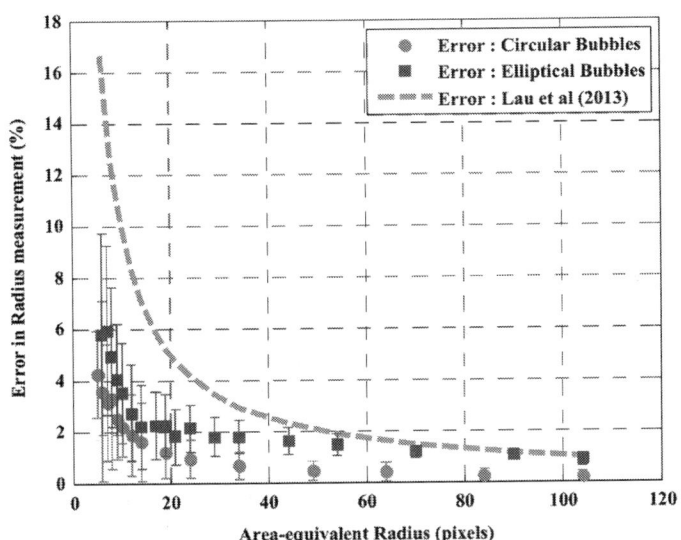

Figure 8: Error in radius measurement of a single bubble for both circular and elliptical bubbles (eccentricity=0.6). Green line shows an estimate of the error which was mathematically obtained by Lau et al. (2013). The dots show the mean error whereas the error bars show the maximum and minimum errors. (For interpretation of the references to color in this figure legend, the reader is referred to the web version of this article.)

For both circular and elliptical bubbles, it was observed that as the area-equivalent radius increased, both the absolute error in the measurement of the bubble size and the uncertainty decreased. The mean error in the size measurement of a single circular bubble decreased from 4.22% at a radius of 5 pixels to about 0.21% at a radius of 104 pixels. The errors in the measurement of an elliptical bubble were found to be slightly greater than the circle with same area-equivalent radius. The mean error in the measurement of an elliptical bubble varied from 5.77% at an area-equivalent radius of 6 pixels to 0.84% at an area-equivalent radius of 104 pixels. For the same area-equivalent radius, the larger uncertainty associated with the elliptical bubble compared to a circular bubble can be attributed to the greater uncertainty involved in the measurement of the minor radius of elliptical bubble, which is smaller than the area-equivalent radius. The uncertainty in the measurement of the bubble size can be caused by many factors such as inhomogeneous background illumination, and background fluctuations due to out-of-focus bubbles etc.

Polydisperse Bubble Size Distribution Measurement

To further validate the effectiveness of the image-processing approach to extract the individual bubbles from bubble clusters, we applied our analysis technique to synthetic bubble images resembling the bubble distributions from our experiments, with average number of bubbles exceeding 500 and void fraction ranging from 0.1 to 0.7. Fig. 9a shows a sample of a synthetic bubble image with 900 bubbles. The bubbles were randomly placed in a 1024×1024 pixel grid and the bubble eccentricity was randomly chosen between 0 and 0.9, which is the range of eccentricity of bubbles observed in our experiments. The results after the image processing are illustrated in Fig. 9b. The red and cyan ellipses denote the in-focus and out-of-focus large bubbles/clusters respectively (the bubbles segmented by the cluster-processing step), the yellow ellipses correspond to the intermediate size bubbles and the magenta dots correspond to tiny spherical bubbles. As the figure shows, all the bubbles are well segmented and the clusters are mostly very well resolved.

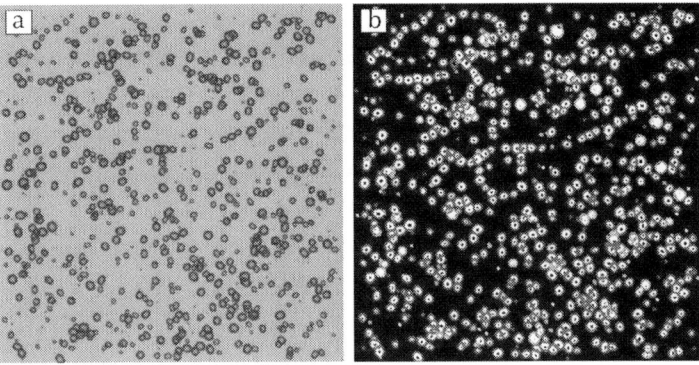

Figure 9: (a) A synthetic bubble image and (b) the corresponding output from the described image-processing algorithm. (For interpretation of the references to color in this figure, the reader is referred to the web version of this article.)

A more quantitative assertion of the efficacy of the current approach is provided by a comparison between the probability density function (PDF) of the actual input bubble size distribution and the PDF obtained after applying the image analysis algorithm to the synthetic bubble image as shown in Fig. 10.

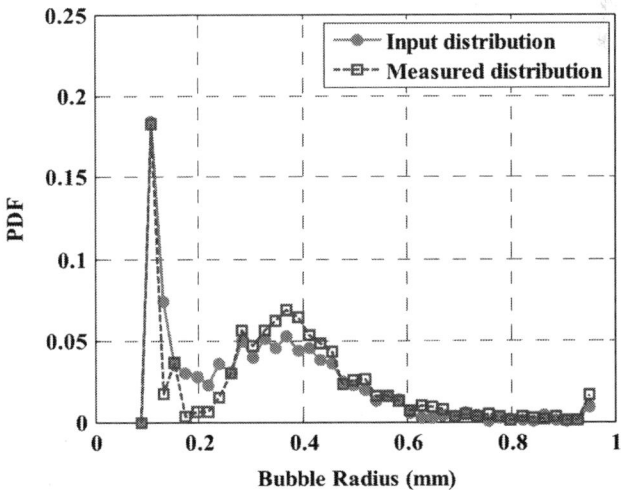

Figure 10: PDFs of bubble size distribution.

The input bubble size distribution was a bimodal distribution with peaks at 0.1 and 0.4 mm, which correspond to a radius of 2 pixels and 8 pixels, respectively. In general, the output PDF from our image analysis showed a good match with the input fraction of bubbles at different sizes, although a slight deviation can be observed for the intermediate bubbles in the bubble radius range of 0.15–0.45 mm. In addition, the image-processing algorithm captures the bimodal distribution quite well and the magnitudes in the PDFs are in good agreement.

A more comprehensive assessment of our approach is provided by implementing it on bubble images generated with different void fractions in the range of 0.02–0.66 and comparing the Sauter mean diameters (d_{32} or SMD) obtained from our analysis with the input. According to Ferreira et al. (2012), the SMD is a significant scale for characterizing mass transfer during the interfacial processes, which is defined as

$$SMD = \frac{\sum_{i=1}^{N} n_i d_i^3}{\sum_{i=1}^{N} n_i d_i^2}$$

where $d_i = \sqrt{a_i b_i}$, a_i and b_i are major and minor axis lengths of the bubble projected area, n_i is the number of bubbles with a particular diameter d_i and N is the total number of bubbles in an image. In general, an increase of SMD corresponds to a decrease of surface-to-volume ratio and implies an increasing portion of larger size bubbles in the image. Hence, as the number of large bubbles in the image increased leading to a change in void fraction from 0.02 to 0.66, the actual SMD increased from 0.21 to 0.71 mm. Bubble images were synthesized with different void fraction values. As the void fraction increased, there was greater overlap between bubbles. For each of these synthesized images, the actual SMD is calculated as the SMD of all the bubbles in the image. Then, the image-processing algorithm was applied and the SMD of all the detected bubbles is calculated. Then, this measured SMD is plotted against the actual SMD for all the images and is shown in Fig. 11. As it shows, the measured SMD is in good agreement with the input results with errors staying within 6–8%. This uncertainty in the measurement can be attributed to the limitation of the image-processing algorithm to accurately resolve highly overlapping bubbles, particularly when individual bubbles completely overlap with clusters, and also to the errors due to non-uniformity in the image background.

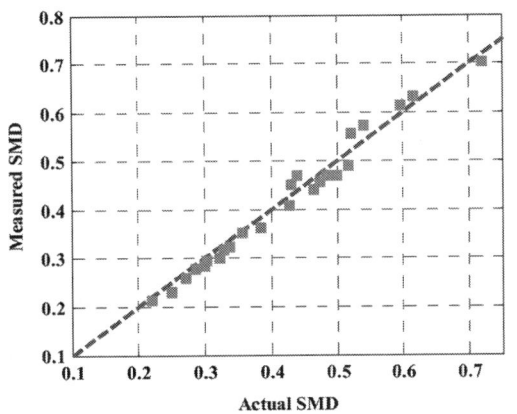

Figure 11: Comparison of actual and measured SMD obtained from synthetic bubble images (The blue line has a slope of 1). (For interpretation of the references to color in this figure legend, the reader is referred to the web version of this article.)

Experimental Validation of the Image-processing Approach

The bubble information extracted from our image analysis can be used to estimate the ventilation flow rate during the experiments, which can be compared with the ventilation rate measured from the mass flow controller to validate the efficacy of our image analysis technique. Note that for this comparison, air absorption in the bubbly wake is neglected considering the small time-scales (<1 s) between air injection and image capture. For such a small time-scale, gas transfer measurements in the bubbly wake indicate a maximum possible increase in dissolved oxygen concentration to be 10^{-5} ppm. Thus, the mass flow rate of air at the two locations can be considered almost equal.

The estimation of ventilation rate from bubble images consists of three steps: (1) extract 2D bubble size information, (2) obtain the volume and velocity of individual bubbles, and (3) calculate the volume flow rate from the volume and velocity information of all the bubbles accounting for the changes in pressure and temperature.

In the first step, the extraction of 2D bubble information is accomplished by the proposed image-processing technique. In this step, the semi-major axis (a) and semi-minor axis (b) lengths of the elliptical bubbles are determined. Second, an elliptical bubble is rotated around its major axis to obtain a prolate spheroid or an ellipsoid of revolution. Then, the volume of this prolate spheroid can be mathematically determined as $(4\varpi/3)b^2a$. Fig. 12 presents the mean streamwise velocity of bubbles in the wake obtained using SIV technique as described in Section 2. As it is shown in Fig. 12, the bubble velocity remains almost uniform in the wake, varying within 5% of about 5 m/s. It is noteworthy that a non-symmetric velocity field is observed with higher velocity above the centerline since the ventilated air is injected from the pressure side of the hydrofoil located below the centerline.

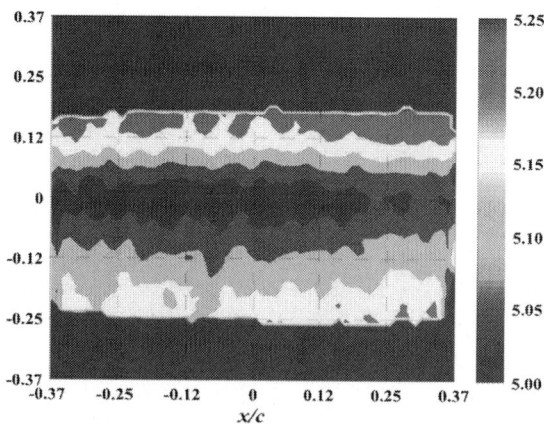

Figure 12: The mean horizontal velocity field obtained using SIV technique (The origin in this contour plot is located at a distance of 377 mm

from hydrofoil center of rotation). The units on the color bar are m/s. (For interpretation of the references to color in this figure legend, the reader is referred to the web version of this article.)

Finally, in the third step, the volume and velocity information of all the bubbles are used to calculate the volume flow rate. To compare with the corresponding volumetric flow rate at standardized temperature and pressure as obtained from the mass flow controller, the volume flow rate measured from bubble images is adjusted with respect to the pressure and temperature in the test-section. It is expected that if the ventilation flow rate is measured at any downstream location in the wake based on the bubble images, the measured ventilation flow rate should match the ventilation rate actually introduced in the flow. In our experiments, the images were captured at three different downstream locations and the ventilation flow rate was measured at all three locations. Finally, steps 1 and 2 are repeated for all the bubbles to obtain a predicted flow rate for one single image. This expression is given by

$$\dot{Q}_{total} = \frac{\beta^3}{M}\Sigma^M_{k=1}\dot{Q}_{single,k}; \ \dot{Q}_{single} = \Sigma^N_{i=1}\left(\frac{u_i}{w}\right)\left(v_i \times \frac{P_{TS}}{P_0} \times \frac{T_0}{T_{TS}}\right);$$

$$v_i = (4\pi/3)a_i b_i^2.$$

where M and N denote number of images over which averaging is done (M=1000 for our experiments) and the total number of bubbles in a single image, respectively.   Represents spatial resolution and w refers to the width of a single bubble image (1024 pixels or 60 mm). u_i and $_i$ signify velocity and volume of individual bubbles, respectively. Finally, P_0 and T_0 correspond to ambient temperature and pressure, and $_{PTS}$ and $_{TTS}$ are corresponding test-section conditions. The volume flow rate is averaged over all the 1000 images to obtain the mean ventilation rate. Fig. 13 below shows the measured mean ventilation rate values at each of the three different downstream locations:

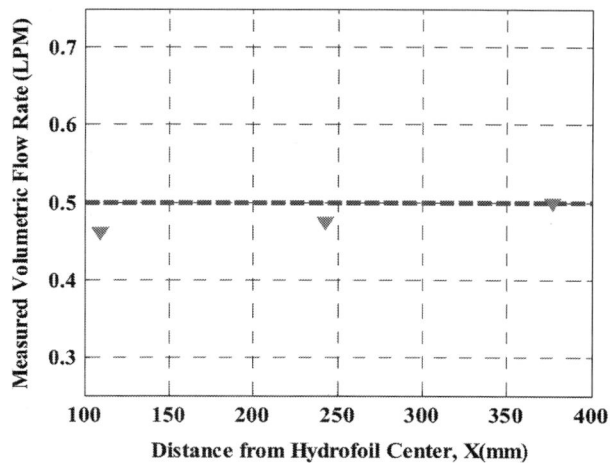

Figure 13: The measured volumetric flow rates as a function of downstream distance in the bubbly wake compared with the input flow rate of 0.5 SLPM (shown in blue). (For interpretation of the references to color in this figure legend, the reader is referred to the web version of this article.)

As the figure above shows, the measured mean ventilation volumetric flow rates compare well with the actual volumetric flow rate of 0.5 SLPM, with the difference within 9% of the actual value at all the three downstream locations. As the clustering phenomena reduces downstream away from the hydrofoil, the mismatch decreases to less than 1%. Larger mismatch occurring closer to the hydrofoil is likely to be associated with the higher uncertainty in extracting bubble information from very dense bubble clusters. The clustering of the bubbles underestimates the volumetric flow rate. This indicates that the accuracy of the image processing technique decreases with excessive clustering.

CONCLUSIONS

In this paper, a robust image measurement technique is proposed to measure the bubble size distribution in dense bubbly flows. The proposed image analysis technique divides all the bubbles into different categories including bubbles within different size range and bubble clusters based on bubble morphology and size. The bubble clusters are passed through an additional step that resolves clusters into individual bubbles. Based on the intensity gradient at the center of individual bubbles, a bubble cluster is divided into in-focus and out-of-focus portions, which are subsequently segmented using 'cluster-processing' and watershed segmentation with distance transform, respectively. We implement the proposed technique to obtain bubble information within a wide size range of 120 μm–4 mm from the images acquired in ventilation experiments for highly turbulent bubbly flows. The image analysis technique was first validated using synthetic bubble images. A further validation was performed by comparing volumetric airflow rates derived according to the bubble size information obtained using the proposed technique with those measured from a mass flow controller.

As the result shows, the image analysis approach is able to capture the size, shape and location information of both in-focus and out-of-focus bubbles from bubble shadow images acquired using SIV technique. In our experiments, the bubble eccentricities largely varied from 0 to 0.9 and the technique was found to be robust enough to accurately resolve the overlapping bubble clusters in this range of bubble eccentricity. Further, the image analysis approach was also applied to simulated bubble images which had some overlapping elliptical bubbles with high eccentricity ($0.95 < e < 1$) and the results proved the efficacy of the image analysis approach in resolving such clusters.

Although the proposed approach was found to work well at a wide range of void fractions (0.02–0.7), it was found that as the bubble clustering reduced, the accuracy of the technique to resolve the bubble clusters increased. Last but not the least, the proposed image analysis technique is computationally efficient. For instance, using an Intel® Core™ i5-3320M (2.60 GHz, 4 GB Memory) processor, the average processing time per image ranged from 3 to 10 s. Therefore, with further development on the parallelization of the algorithm over multiple cores, the proposed technique will allow real-time monitoring of bubble distribution for a wide range of industrial applications. The proposed technique can be broadly applied to bubble characterization in different experimental facilities (e.g. inclined, vertical channels, etc.), as well as other types of two-phase flows such as sprays, drops and froth to quantify void fraction, gas hold-up and size distribution of the dispersed phases.

ACKNOWLEDGMENTS

The authors gratefully acknowledge the support by U.S. Department of Energy (Grant No. DE-EE0005416), ALSTOM Renewable Power Canada Inc. and the Office of Naval Research (Program manager, Dr. Ronald Joslin, Grant No. N000140910141). Help extended by undergraduate student Mr. Siyao Shao during the course of this work is also acknowledged and appreciated. Thanks are also due to Mr. Garrett Monson for his gas transfer measurements.

REFERENCES

1. Adrian, R.J., 1991. Particle-imaging techniques for experimental fluid mechanics. Annu. Rev. Fluid Mech. 23 (1), 261–304. http://dx.doi.org/10.1146/annurev. fl.23.010191.001401.

2. Bailey, M., Gomez, C., Finch, J., 2005. Development and application of an image analysis method for wide bubble size distributions. Miner. Eng. 18 (12), 1214–1221. http://dx.doi. org/10.1016/j.mineng.2005.07.019.

3. Bonifazi, G., Serranti, S., Volpe, F., Zuco, R., 1999. A combined morphological and color based approach to characterize flotation froth bubbles. Proc. Second Int. Conf. Intell. Process. Manuf. Mater. 1, 465–470.

4. Bröder, D., Sommerfeld, M., 2007. Planar shadow image velocimetry for the analysis of the hydrodynamics in bubbly flows. Meas. Sci. Technol. 18 (8), 2513–2528. http://dx.doi. org/10.1088/0957-0233/18/8/028.

5.   Chen, Q., Yang, X., Petriu, E.M., 2004. Watershed segmentation for binary images with different distance transforms. In: Proceedings of the 3rd IEEE International Workshop on Haptic, Audio and Visual Environments and their Applications, pp. 111–116.

6.   do Amaral, C.E.F., Alves, R., da Silva, M.J., Arruda, L.V.R., Dorini, L., Morales, R.E.M., Pipa, D.R., 2013. Image processing techniques for high-speed videometry in horizontal two-phase slug flows. Flow Meas. Instrum. 33, 257–264. http://dx. doi.org/10.1016/j.flowmeasinst.2013.07.006.

7.   Ferreira, A., Pereira, G., Teixeira, J.A., Rocha, F., 2012. Statistical tool combined with image analysis to characterize hydrodynamics and mass transfer in a bubble column. Chem. Eng. J. 180, 216–228. http://dx.doi.org/10.1016/j. cej.2011.09.117.

8.   Geary, N.W., Rice, R.G., 1991. Bubble size prediction for rigid and flexible spargers. AIChE J. 37, 161–168. http:// dx.doi.org/10.1002/aic.690370202.

9.   Glover, A., Skippon, S., Boyle, R., 1995. Interferometric laser imaging for droplet sizing, a method for droplet-size measurement in sparse spray systems. Appl. Opt. 34 (36), 8409–8421. http://dx.doi.org/10.1364/AO.34.008409.

10. Eddins, S.L., Gonzalez, R.C., Woods, R.E., 2004. Digital Image Processing Using Matlab. Princeton Hall Pearson Education Inc., New Jersey.

11. Goss, L.P., Estevadeoral, J., Crafton, J.W., 2007. Velocity measurements near walls, cavities, and model surfaces using Particle Shadow Velocimetry (PSV). In: Proceedings of the 22nd International Congress on Instrumentation in Aerospace Simulation Facilities.

12. Grant, I., 1997. Particle image velocimetry: a review. Proc. Inst. Mech. Eng. Part C 211 (1), 55–76. http://dx.doi. org/10.1243/0954406971521665.

13. Haralick, R.M., Shapiro, L.G., 1992. Computer and Robot Vision. Addison-Wesley, New York

14.   Honkanen, M., Eloranta, H., Saarenrinne, P., 2010. Digital imaging measurement of dense multiphase flows in industrial processes. Flow Meas. Instrum. 21 (1), 25–32. http://dx.doi.org/10.1016/j.flowmeasinst.2009.11.001.

15.   Honkanen, M., Saarenrinne, P., Stoor, T., Niinimaki, J., 2005. Recognition of highly overlapping ellipse-like bubble images. Meas. Sci. Technol. 16, 1760–1770. http: //dx.doi.org/10.1088/0957-0233/16/9/007.

16.   Junker, B., 2006. Measurement of bubble and pellet size distributions, past and current image analysis technology. Bioproc. Biosyst. Eng. 29 (3), 185–206. http: //dx.doi.org/10.1007/s00449-006-0070-3.

17.   Kamp, A., Chesters, A., Colin, C., Fabre, J., 2001. Bubble coalescence in turbulent flows, a mechanistic model for turbulence-induced coalescence applied to microgravity bubbly pipe flow. Int. J. Multiph. Flow 27 (8), 1363–1396. http: //dx.doi.org/10.1016/S0301-9322(01)00010-6.

18.   Kracht, W., Emery, X., Paredes, C., 2013. A stochastic approach for measuring bubble size distribution via image analysis. Int. J. Miner. Process 2013 (121), 6–11. http: //dx.doi.org/10.1016/j.minpro.2013.02.016.

19.   Laakkonen, M., Moilanen, P., Miettinen, T., Saari, K., Honkanen, M., Saarenrinne, P., 2005. Local bubble size distributions in agitated vessel: comparison of three experimental techniques. Chem. Eng. Res. Des. 83 (1), 50–58. http://dx.doi.org/ 10.1205/cherd.04122.

20.   Lau, Y.M., Deen, N.G., Kuipers, J.A.M., 2013. Development of an image measurement technique for size distribution in dense bubbly flows. Chem. Eng. Sci. 94, 20–29. http://dx.doi.org/10.1016/j.ces.2013.02.043

21.   Lin, B., Recke, B., Knudsen, J.K., Jørgensen, S.B., 2008. Bubble size estimation for flotation processes. Miner. Eng. 21 (7), 539–548. http://dx.doi.org/10.1016/j. mineng.2007.11.004.

22.   Liu, T., Bankoff, S., 1993. Structure of air–water bubbly flow in a vertical pipe – II. Void fraction, bubble velocity and bubble

size distribution. Int. J. Heat Mass Transf. 36 (4), 1061–1072. http://dx.doi.org/10.1016/S0017-9310(05)80290-X.

23. Mena, P.C., Pons, M.N., Teixeira, J.A., Rocha, F.A., 2005. Using image analysis in the study of multiphase gas absorption. Chem. Eng. Sci. 60 (18), 5144–5150. http: // dx.doi.org/10.1016/j.ces.2005.04.049.

24. Meyer, F., 1994. Topographic distance and watershed lines. Signal Process 38 (1), 113–125. http://dx.doi. org/10.1016/0165-1684(94)90060-4.

25. Mudde, R., Groen, J., Van Den Akker, H.E.A., 1998. Application of LDA to bubbly flows. Nucl. Eng. Des. 184 (2–3), 329–338. http://dx.doi.org/10.1016/S0029- 5493(98)00206-4.

26. Otsu, N., 1979. A threshold selection method from gray-level histograms. IEEE Trans. Syst. Man Cybern. SMC-9, 62–66.

27. Pla, F., 1996. Recognition of partial circular shapes from segmented contours. Comput. Vis. Image Underst. 63 (2), 334–343. http://dx.doi.org/10.1006/ cviu.1996.0023

28. Prasser, H., 2008. Novel experimental measuring techniques required to provide data for CFD validation. Nucl. Eng. Des. 238, 744–770. http://dx.doi.org/10.1016/ j.nucengdes.2007.02.050.

29. Roesler, T., Lefebvre, A., 1989. Studies on aerated-liquid atomization. Int. J. Turbo Jet-Engines 6 (3–4), 221–230. http:// dx.doi.org/10.1515/TJJ.1989.6.3-4.221

30. Saberi, S., Shakourzadeh, K., Bastoul, D., Militzer, J., 1995. Bubble size and velocity measurement in gas–liquid systems: application of fiber optic technique to pilot plant scale. Can. J. Chem. Eng. 73 (2), 253–257. http://dx.doi.org/10.1002/ cjce.5450730213.

31. Sadr-Kazemi, N., Cilliers, J., 1997. An image processing algorithm for measurement of flotation froth bubble size and shape distributions. Miner. Eng. 10 (10), 1075–1083. http:// dx.doi.org/10.1016/S0892-6875(97)00094-0.

32. Sahoo, P.K., Arora, G., 2004. A thresholding method based on two-dimensional Renyi's entropy. Pattern Recognit. 37 (6), 1149–1161. http://dx.doi.org/10.1016/j.patcog.2003.10.008.

33. Sahoo, P., Wilkins, C., Yeager, J., 1997. Threshold selection using Renyi's entropy. Pattern Recognit. 30 (1), 71–84. http://dx.doi.org/10.1016/S0031-3203(96) 00065-9.

34. Smith, J.S., Burns, L.F., Valsaraj, K.T., Thibodeaux, L.J., 1996. Bubble column reactors for wastewater treatment. 2. The effect of sparger design on sublation column hydrodynamics in the homogeneous flow regime. Ind. Eng. Chem. Res. 35 (5), 1700–1710. http://dx.doi.org/10.1021/ie950366y.

35. Soille, P., 2003. Morphological Image Analysis: Principles and Applications, 2nd ed. Springer-Verlag, New York.

36. Tayali, N., Bates, C., 1990. Particle sizing techniques in multiphase flows: a review. Flow Meas. Instrum. 1 (2), 77–105. http://dx.doi.org/10.1016/0955-5986(90) 90032-3.

37. Zaidi, S.H., 1998. Difficulties in measuring liquid droplet size distributions using laser diffraction technique. Atomization Spray 8 (4), 439–452

38. Zhang, G., Hong, Z., Ning, X., 2011. Flotation bubble image segmentation based on seed region boundary growing. Min. Sci. Technol. (China) 21 (2), 239–242. http: //dx.doi. org/10.1016/j.mstc.2011.02.013.

39. Zhou, K., Yang, C., Gui, W., Xu, C., 2010. Clustering-driven watershed adaptive segmentation of bubble image. J. Cent. South Univ. T 17 (5), 1049–1057. http: //dx.doi.org/10.1007/ s11771-010-0597-y.

Multiphase Monolith Reactors: Chemical Reaction Engineering of Segmented Flow in Microchannels

Michiel T. Kreutzer[a], Freek Kapteijn[a], Jacob A. Moulijn[a], Jand ohan J. Heiszwolf[b]

[a]Delft University of Technology, Reactor and Catalysis Engineering, Julianalaan 136, 2628 BL Delft, The Netherlands
[b]Albemarle Catalyst Company BV, Nieuwendammerkade 1–3, 1030 BE, Amsterdam, The Netherlands

ABSTRACT

The use of segmented flow in capillaries, also known as Taylor flow, for reaction engineering purposes has soared in recent years. On the one hand, Taylor flow has been used in honeycomb monolith catalyst supports. On the other hand, Taylor flow is the common flow pattern in multiphase micro channel reactors. This contribution reviews the fluid mechanical aspects of this flow pattern in quite general terms, with an emphasis on the underlying principles. From very simple analysis, design estimates for mass transfer, pressure drop and residence time distribution may be obtained with relative ease and—for multiphase reactors—surprising accuracy.

INTRODUCTION

Monolith reactors are attracting more and more attention as alternatives for both three-phase slurry reactors (e.g. Cybulski et al., 1999, Broekhuis et al., 2001, Machado et al., 1999, Boger et al., 2003 and Boger et al., 2004a) and trickle bed reactors (e.g. Nijhuis et al., 2003a and Edvinsson and Cybulski, 1995). The operating mode depends on the size of the straight parallel channels: in large channels the fluid trickles downwards along the channel walls and the gas travels, co-currently or countercurrently, through the channel in the core. In smaller channels, the dominant flow pattern is a segmented slug flow or bubble-train flow of elongated bubbles and slugs. In the beginning of this paper, criteria to predict the different multiphase flow regimes will be briefly reviewed, and the remainder of the paper deals with the segmented flow pattern. For the trickle or film flow pattern, the interested reader is referred to Lebens (1999) and Heibel et al., 2001, Heibel et al., 2002, Heibel et al., 2004a and Heibel et al., 2004b.

The surface tension dominated flow of elongated bubbles (Olbricht, 1996 and Chang, 2002) has been recognised as a useful flow pattern for various applications outside of chemical reactor engineering. Perhaps the simplest and oldest one is the use of a bubble as a flow meter (Fairbrother and Stubbs, 1935). Because the bubble extends over almost the entire cross-sectional area of the channel, the velocity of the bubble is nearly equal to the velocity of the liquid upstream and downstream of the bubble, and by visual observation the velocity of the bubble can easily be measured.

Other applications exploit the enhanced mass or heat transfer due to circulation, e.g. the improvement of microfiltration by adding gas bubbles to the capillary channels (Laborie et al., 1998 and Laborie et al., 1999). The experimentally observed enhancement of microfiltration efficiency is also attributed to the removal of filter cake by the pressure pulsing caused by the passing of the bubbles and slugs (Cui et al., 2003). Circulation in the plasma separating red blood cells in microvascular flow might enhance oxygen uptake and release, but the evidence for such convective enhancement is incomplete (Bos et al., 1996). Finally, the liquid slugs are practically sealed between two bubbles. This feature was used to make so-called continuous flow analysers (Thiers et al., 1971), with Technicon's autoanalyser as the best known brand name. In these machines, the samples that must be analysed are injected into a capillary and separated by bubbles. Since the bubbles prevent mixing of the samples, long capillary tubes with multiple analysis sections can be used with minimal mixing of consecutive samples. So, in fact the autoanalyser is an example of high throughput analysis *avant la lettre*.

In this paper, we consider the use of capillary channels for three-phase reactions, i.e., the reactions of a gas component with a liquid component on a solid catalyst surface. As can be deduced from the examples above, the bubble train flow pattern has many useful features: plug flow (no macromixing) combined with good mass transfer (local mixing) and low pressure drop for a given specific contact area. The aim of this paper is to review the fluid mechanics and transport phenomena of the Taylor flow pattern. The literature on this subject is very scattered, in part due to the very different areas where it is applied, from medical phenomena such as lung opening (Grotberg, 2001 and Bilek et al., 2003) and sample analysis to nuclear reactor cooling and membrane filtration.

While much of the work in the field of chemical engineering was carried out for monolith honeycomb structures with channels of about 1 mm diameter, the results are with minor adjustments applicable to multiphase microreactors, which typically have slightly smaller channels of about 300m.

Several chemical aspects play a role. A suitable catalyst must be applied to the wall. The preparation of such catalyst is beyond the scope of this paper, but we would like to point the interested reader to the preparation reviews by Nijhuis et al. (2001a) and Vergunst et al. (2001). Apart from a short overview of multiphase processes that are currently considered or carried out in capillary channels, we will not consider the chemistry in any detail.

AREAS OF APPLICATION

Monoliths or honeycombs (the latter term is more common in the USA) were originally designed for single phase use in the automotive catalyst converter. In the 1970s Horvath et al. (1973) used the slug flow pattern to enhance the mass transfer in a liquid–solid process, where the only role of the gas was to enhance the mass transfer to an immobilised enzyme on the channel wall. Pioneering use of monoliths was performed in the 1980s at Chalmers, where the hydrogenation of nitroaromatics (Hatziantoniou and Andersson, 1984 and Hatziantoniou et al., 1986) was performed. The thesis works of Irandoust (1989) andEdvinsson (1994) give a good overview of the reactions performed: the hydrogenation of 2-ethyl-hexenal (Irandoust and Andersson, 1988), anthraquinones (Irandoust et al., 1989) and acetylene, and the hydrogenolysis of thiophene and dibenzothiophene (Edvinsson and Irandoust, 1993).

Akzo-Nobel now operates a reactor in the anthraquinone process of hydrogen peroxide production (Edvinsson Albers et al., 2001). This process is a two-step loop process. Carry-over of catalyst from one step of the process to the next must be prevented, so slurry catalysts are less attractive. The hydrogenation is fast, and the good mass transfer of the capillares make the honeycombs a suitable fixed-bed reactor.

In our department, the hydrogenation reactions of styrene (Smits et al., 1996 and Nijhuis et al., 2003a), -methylstyrene (Kreutzer et al., 2001), cinnemaldehyde, glucose, benzaldehyde (Nijhuis et al., 2001b), 3-methyl-1-pentyn-3-ol (Nijhuis et al., 2003b) and edible fats (Boger et al., 2004b) have been performed, as well as oxidations (Crezee et al., 2001), Fischer–Tropsch synthesis (de Deugd et al., 2003) and enzymatic reactions (De Lathouder et al., 2004).

Klinghoffer et al. (1998) used up-flow monoliths for the wet air oxidation of acetic acid. Winterbottom and co-workers (Winterbottom et al., 2003, Natividad et al., 2004 and Marwan and Winterbottom, 2004) studied the hydrogenation of 2-butyne-1,4-diol in downflow monoliths. Haakana et al. (2004) oxidised -D-lactose to D-lactobionic acid sodium salt in a monolith pilot reactor. Biochemical applications include bio-degradations (Quan et al., 2003) and biomass upgrading (Schutt et al., 2002).

Air Products and Chemicals, Inc. have patented the use of monoliths for fast and highly exothermic nitroaromatic hydrogenations (Machado et al., 1999). This design, which in many ways is similar to venturi loop reactors, exploits the very low pressure drop in monoliths: a safe reactor design requires the product to be recycled through the reactor several tens or several hundreds of times, and the pumping energy for the recycle can be reduced significantly with a low-pressure-drop monolith. Apart from several nitro-aromatics (Broekhuis et al., 2001), also sorbitol hydrogenation was considered (Broekhuis et al., 2004).

Other efforts in industry include olefin hydrogenation and toluene saturation by Corning, Inc (Liu, 2002) and a post-reactor in the production of phtalic anhydride (Eberle et al., 2000).

In short, monoliths have been applied for a wide range of processes. Fast multiphase reactions, typically hydrogenations, seem to benefit the most from the application of monoliths. In microreactor technology, the use of segmented flow is more recent (e.g. Song et al., 2004). On microfabricated integrated systems, much more control of the inlet and outlet geometries is possible, and the research focuses on general features of the integrated microfluid system, such as capillary gas–liquid separators downstream the reaction zone (Günther et al., 2004). An elegant application that uses the plug flow characteristic is the production of colloidal silica (Khan et al., 2004).

TWO-PHASE FLOW PATTERNS IN CAPILLARY CHANNELS

Defining Small Channels

Two-phase flow through capillary channels is considerably different from two-phase flow through larger channels: viscous ($\sim \mu u/d$) and interfacial ($\sim \gamma/d$) stresses, both inversely proportional to the diameter, are more important than inertial ($\sim \rho u^2$) and gravitational ($\sim \rho g H$) stresses.

A reasonable definition of the term *capillary* might be obtained by requiring the dominance of surface tension forces over buoyancy. It has been shown analytically (Bretherton, 1961) that the rise velocity of an elongated bubble in a sealed liquid phase capillary vanishes for

$$Bo = \frac{\rho g d^2}{\gamma} < 3.368$$

(1)

For water–air, this implies that a capillary has d<5 mm, while for the decane-air we have d<3.4 mm. This definition of a threshold diameter has the advantage of taking the effect of some fluid properties into account, but is still imperfect: it lacks viscosity as a parameter and for non-axisymmetric duct geometries, the constant in Eq. (1) changes. On the other hand, experimental evidence suggests that significant deviations from large tubes occur in air-water systems at d≈5 mm, which suggests that the threshold defined by Eq. (1) is appropriate.

Observed Flow Patterns

Many flow patterns have been described for two-phase flow in capillaries. Although objective methods, e.g. based on void fraction measurements (Dowe and Rezkallah, 1999) are under development, this is usually done by visual observation, and the discrimination of the different patterns is rather subjective. Most researchers' present "representative" pictures of the observed flow pattern for clarity, see Fig. 1 for a schematic example.

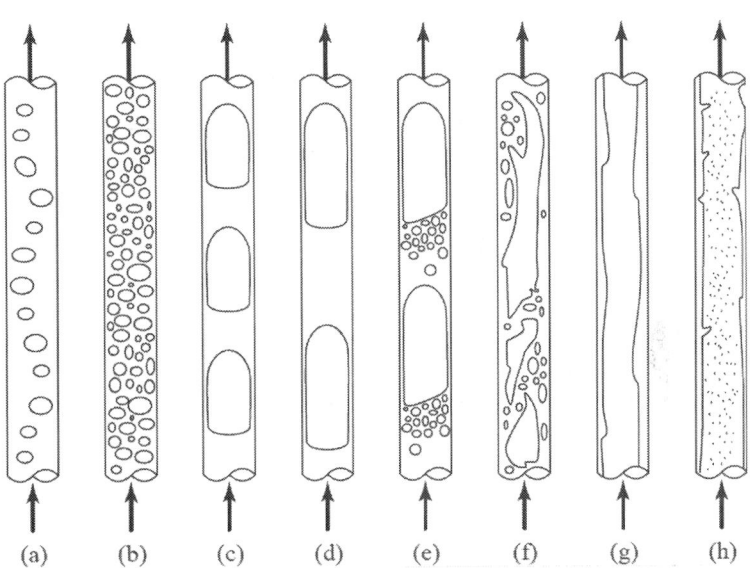

(a)   (b)   (c)   (d)   (e)   (f)   (g)   (h)

Figure 1: Sketch of observed flow patterns in capillary channels. (a, b): bubbly flow, (c, d) segmented flow (a.k.a. bubble train flow, Taylor flow, capillary slug flow), (e) transitional slug/churn flow, (f) churn flow, (g) film flow (down flow only), (h) annular flow.

Typically—with the aforementioned limitations in mind—the number of different flow patterns may be reduced to the following five.

- At very low liquid superficial velocities, typically in the order of a few mm/s, a *film flow* pattern is feasible, in which the liquid flows downwards on the walls of the channel and the gas flows through the centre, either upward or downward.
- In *bubbly flow*, the non-wetting gas flows as small bubbles dispersed in the continuous, wetting liquid. This pattern is observed for low gas fractions at moderate velocities, i.e., those conditions in which coalescence is minimal.

- *Taylor flow*, sometimes called *plug flow, slug flow, bubble train flow, segmented flow* or *intermittent flow* is the flow pattern of large long bubbles that span most of the cross-section of the channel. The relevant lengths are mainly determined by the inlet conditions, as was demonstrated by Kreutzer et al. (2003), who were able to obtain a wide variety of bubble and slug lengths by varying the inlet conditions alone.

- At higher velocities, small satellite bubbles appear at the rear of the slug, and the slugs eventually are aerated (Reinecke and Mewes, 1999). The chaotic flow pattern that emerges when the velocity is increased even further is called *churn flow*.

- At high velocities and low liquid fraction, the *annular flow* pattern exists of a thin wavy liquid film flowing along the wall with a mist of gas and entrained liquid in the core.

Flow Transitions

Typically, so–called flow maps are constructed of liquid superficial velocity versus gas superficial velocity. In these maps, experimentally determined flow patterns are plotted with distinct markers, and the boundaries, i.e., the transitions of one flow pattern to the other, are plotted by lines.

Which flow pattern actually occurs in a given capillary channel depends on the gas and liquid properties (ρ_G, μ_G, ρ_L, μ_L, γ), duct geometry (at least d) and gas and liquid velocities (u_{Ls}, u_{Gs}). The number of relevant dimensionless groups is large, and most experimental flow maps in the literature are applicable only to the specific systems in which they were obtained. Most of the transitions depend on a disturbance to grow, and the amplitude of the disturbances introduced has a profound effect on the flow map. This was already pointed out by Satterfield and Özel (1977), who reported that the boundary between falling film flow and slug flow is in fact a broad region where the observed flow pattern depends on the method of introduction. Galbiati and Andreini (1992) demonstrated that a smooth introduction of gas and liquid into the capillary channel resulted in stratified and dispersed flow. By introducing only a single thin wire into the water feed of the channel, these flow patterns vanished completely and only slug flow and annular flow were observed. The same effect was observed by using an extremely long calming section that allowed even the smallest disturbances to grow.

While most flowmaps are presented without attempting to account for the effect of fluid properties and channel diameter, some noticeable exceptions exist. Suo and Griffith (1964) performed experiments using octane, heptane and water as liquids and helium, nitrogen and argon as gases. No significant changes were found for the different gases and the groups (ρ_G/ρ_L) and (μ_G/μ_L) were eliminated from consideration. A transition from slug flow to churn flow was given by Re We$=2.8\times10^5$, which agrees more or less with aeration of the slugs at the development of turbulence, see Fig. 2.

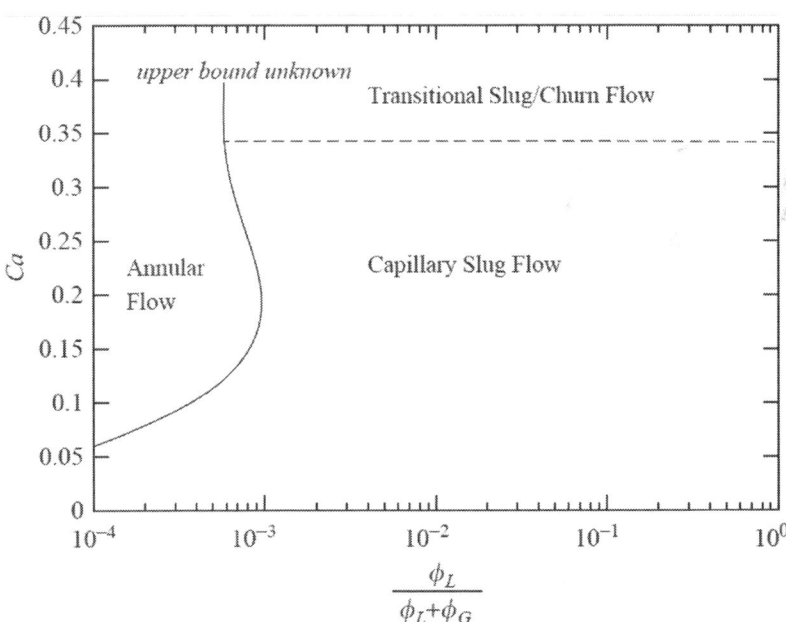

Figure.2: Flowmap of Suo and Griffith for Ca /Re$=1.5\times10^{-5}$. Note that the line separating transitional flow and slug flow coincides with Re$=2200$ (after Suo and Griffith, 1964).

Zhao and Rezkallah (1993) (later updated with new literature data by Rezkallah (1996)) showed that three distinct regimes may be identified: (1) a surface tension dominated regime with bubbly and slug flow, (2) an inertia dominated regime with annular flow and (3) a transitional regime in between with churny flows. Then, the boundary between the regimes is determined by the Weber number, which they based on gas properties and gas superficial velocity

$We_{Gs} = \rho_G \bar{u}_{Gs} d / \gamma$). Roughly, the surface tension dominated regime was delimited by $We_{Gs} < 1$ and the inertial regime was delimited by $We_{Gs} > 20$.

Jayawardena et al. (1997) extended the models of Rezkallah by incorporating viscous effects, based on microgravity experiments alone. In a plot of (Re_{Gs}/Re_{Ls}) versus ((Re_{Ls}/Ca)Re_{Ls}/Ca), the boundaries for a large set of experimental data, obtained using different fluids and geometries, could be accurately predicted, seeFig. 3(a-b).

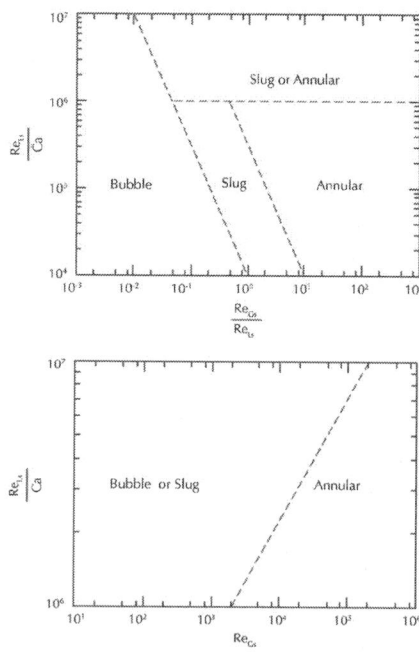

Figure 3: Flowmaps from Jayawardena et al. (after Jayawardena et al., 1997).

For raw experimental data in dimensional form the interested reader is referred to Satterfield and Özel (1977), Triplett et al. (1999), Fukano and Kariyasaki (1993), Mishima and Hibiki (1996), Akbar et al. (2003),Wölk et al. (2000), Günther et al. (2004), Cubaud and Ho (2004).

FUNDAMENTALS OF ELONGATED BUBBLES IN CAPILLARIES

The understanding of segmented flow in capillary channels begins with two now classical papers from the Cavendish Laboratories in Cambridge, published in 1961. In the first paper, Taylor (1961) postulated the main features of the possible flow patterns, introducing the recirculating vortex in the slug, separated from the film attached to the wall (Fig. 4). In later studies, these features of *Taylor* flow were confirmed using photographs and particle image velocimetry (Cox, 1963, Cox, 1964, Thulasidas et al., 1997 and Günther et al., 2004).

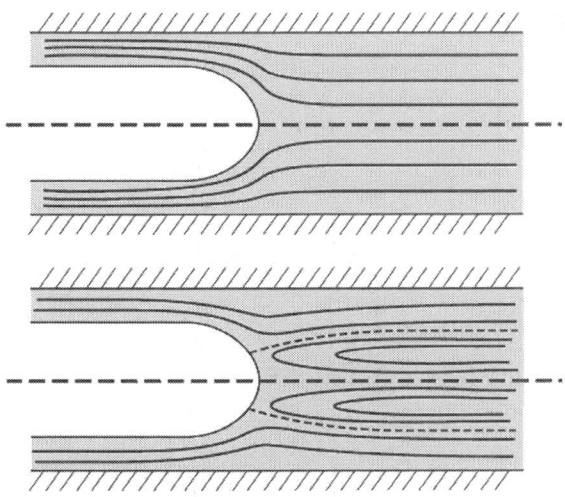

Figure 4: Sketch of the possible liquid streamlines for the flow of elongated bubbles in capillaries. The top pattern occurs for r Ca > 0.7and is

called complete bypass flow. The bottom part, with the stagnation ring on the nose of the bubble, occurs for Ca < 0.7 (after Taylor, 1961).

Lubrication Analysis of Viscous and Interfacial Stresses

The second paper, by Bretherton (1961), pioneered the use of a lubrication analysis for the transitional region where the film is formed, i.e., between the spherical front of the bubble and the flat film far behind the front (Fig. 5). Lubrication theory was originally developed to explain why no solid-to-solid contact occurs in bearings due to the motion of a lubricating viscous fluid. Although in Taylor flow the thin film prohibits gas-to-solid contact rather than solid-to-solid contact, the same mathematical treatment of the equations of fluid motion as developed for the bearings by Reynolds (1886) may be used.

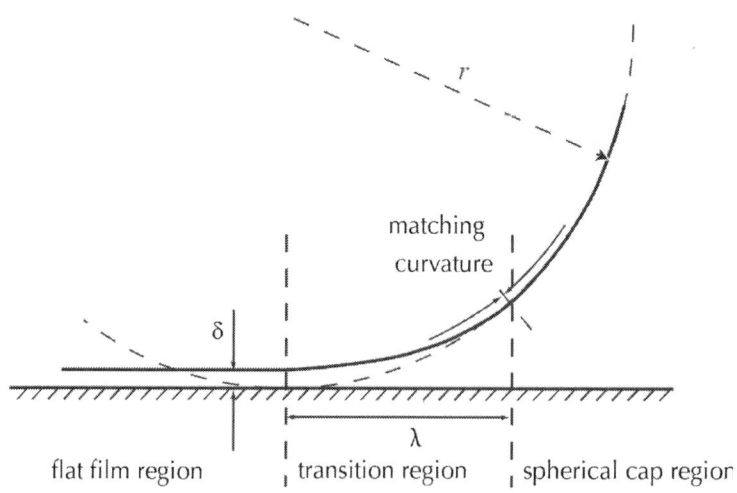

Figure 5: Schematic representation of the transition region between the flat film and the spherical front of the bubble (after Bretherton, 1961).

The full analysis of Bretherton is lengthy, but thorough. Here only a condensed scaling analysis fromAussilous and Quére (2000) is given. The front of the bubble may be regarded as spherical with radius r , so the Laplace pressure difference across the gas–liquid interface is given by $\Delta p = 2\gamma/r$, provided the film thickness is small ($\delta \ll r$). In the region of constant film thickness, the curvature in the axial direction vanishes, and the Laplace pressure difference is given by $\Delta p = \gamma/r$. A balance of the viscous force and the pressure gradient in the transitional region yields

$$\frac{\mu u}{\delta^2} \sim \frac{1}{\lambda}\frac{\gamma}{r},$$

(2)

where λ is the length of the transitional region between the spherical and flat interface. The length λ is unknown, but we can estimate it by requiring that the Laplace pressure is continuous at the interface or, in other words, that the curvature of the spherical part matches the curvature at the end of the transition region.

$$-\frac{\gamma}{r} - \frac{\gamma\delta}{\lambda^2} \sim -\frac{2\gamma}{r},$$

(3)

or $\lambda = \sqrt{\delta r}$, which yields upon substitution into Eq. (2) the now classical scaling rule $\delta/r \, Ca^{2/3}$. The more rigorous analysis of the full Navier–Stokes equations in the transition region at the front and the back of the bubble by Bretherton results in

$$\frac{\delta}{d} = 0.66 Ca^{2/3}$$

(4)

The method developed by Bretherton also gave an expression for the pressure drop over the bubble. In fact, the excess bubble velocity, film thickness and pressure drop are all related.

$$\frac{\Delta p}{\gamma/d} = 7.16(3Ca)^{2/3}$$

(5)

The theory of Bretherton agrees nicely with experimental data for low Capillary numbers. Fig. 6 shows how well the theoretical predictions, which contain no fitted parameters, agreed with experiment. However, for $Ca < 10^{-4}$ and for $Ca > 10^{-2}$, the Bretherton analysis breaks down. We will first review which additional stresses should be considered to explain the experimental observations. Later, we will address the more practical body of literature that has tried to provide engineering correlations that capture deviations from lubrication.

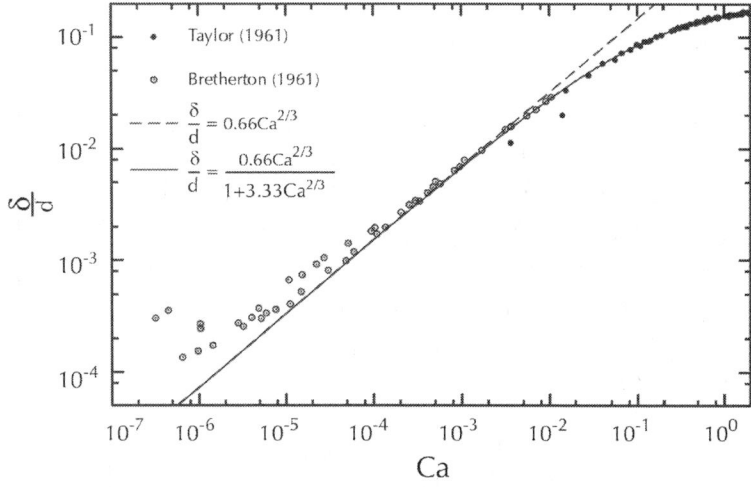

Figure 6: Film thickness between elongated bubbles and a capillary wall, as a function of the capillary number Ca. Bretherton used aniline and benzene, while Taylor used a syrup–water mixture, glycerin and a lubricating oil. (From Bretherton, 1961 and Taylor, 1961.)

Marangoni Effects

As Taylor flow is dominated by surface tension effects, it is no surprise that small amounts of contaminants have a huge impact. Impurities on the gas–liquid interface change the surface tension and gradients of the concentration of these impurities lead to a gradient in the interface pressure. The usual no-shear boundary condition for the gas–liquid interface then breaks down. Ratulowski and Chang (1990) showed that taking into account these surfactant effects, also known as Marangoni effects, explains the difference between theory and experiment for very dominant surface tension (low *Ca*).

The largest concentration gradients are found near the stagnation rings on the bubble, see Fig. 7. At the nose, surfactants are swept from this ring to the axis of the channel or to the film separating the bubble and the wall. On the ring at the rear, surfactants are desorbed to the bulk liquid. The maximum impact of the Marangoni effects may be modelled by assuming no-slip on the interface. Bretherton (1961) showed that using no-slip in the transition regions increases the film thickness and the pressure drop over the bubble by a factor $4^{2/3}$. Most of the experimental on film thickness is indeed bounded by Brethertons "clean" and "hard" limits. The full solution of the interplay of bulk diffusion, surface adsorption and convection of surfactants is much more complex, and the interested reader is referred to the works of Ratulowski and Chang (1990), Park (1992), Stebe and Barthes-Biesel (1995), Waters and Grotberg (2002), Severino et al. (2003) and Ghadiali and Gaver III (2003).

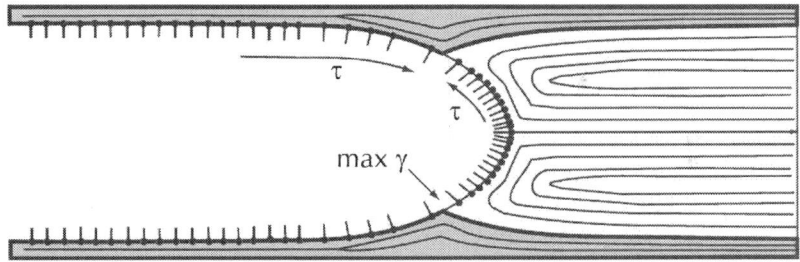

Figure 7: Schematic representation of the surface concentrations and the interaction with the fluid flow (after Ratulowski and Chang, 1989).

Inertial Effects

In experimental work the Capillary number is usually varied by increasing the viscosity of the liquid, and keeping the velocity low for ease of observation. As a consequence, most of the experimental data was obtained at low Reynolds numbers for which simplification to Stokes flow seemed justifiable.

The effects of inertia may in a first analysis be ignored in small capillaries. Bretherton proposed $We \ll 1$ as a suitable criterion. In monolith channels, where $Ca = O\ (10^{-2})$ and $Re = O\ (10^{2})$, the Weber number ($We = Ca\ Re$) readily approaches unity and inertia should probably be considered.

Numerically, the effect of inertia was first included by Edvinsson and Irandoust (1996). They found that the rear spherical caps were flattened and the amplitude of the ripples increased as the Reynolds number increased. Giavedoni and Saita, 1997 and Giavedoni and Saita, 1999 found a slight decrease in film thickness up to $Re=100$. Heil (2001) reported noticeable changes in film thickness and pressure drop when inertia was taken into account in a numerical analysis for Reynolds numbers up to 280. Kreutzer et al. (2005b) simulated entire bubbles and slugs for Re up to 900, using a hybrid upwinding scheme that introduced minimal amounts of numerical diffusion.

All numerical investigations agree that the film thickness decreases slightly from $Re=0$ to 10^{2}, and from then on increases monotonically with increasing Re. Experimental data at high Reynolds numbers were recently reported by Aussilous and Quére (2000), who measured film thickness at high velocities using low viscosity liquids. A noticeable increase in film thickness was found and explained using a scaling analysis, see Fig. 8.

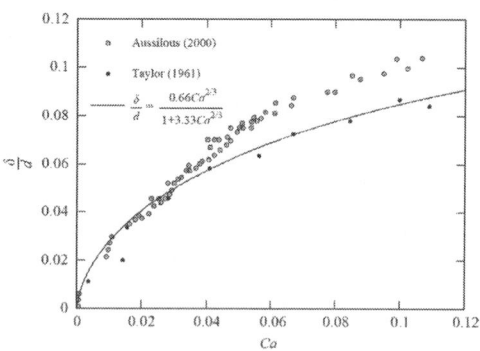

Figure 8: Effect of liquid viscosity on the film thickness between elongated bubbles and a capillary wall, as a function of the capillary numberCa

. The data from Taylor (1961) were obtained with viscous fluids, the data from Aussilous and Quére (2000) were obtained with ethanol. (reproduced with permission from Aussilous and Quére, 2000, copyright 2000, American Institute of Physics).

Gravitational Effects

The effect of gravity in Bretherton's problem is very limited. Hazel and Heil (2002) studied the impact of gravity on the film thickness and pressure drop and found only a small difference for Bo = ±0.43 (the sign of the Bond number indicates up-flow or downflow). Edvinsson and Irandoust (1996) performed several calculations for Re=200 and -2<Fr<2. The effect of gravity on the film thickness was significant for Bo > 1. In square capillaries, the influence of gravity on the film thickness is more pronounced than in circular capillaries: for up-flow smaller bubble radii are reported, and the reverse effect was observed experimentally and numerically for downflow.

Numerical Analysis

The theory of Bretherton has been confirmed by numerous numerical studies, in which, of course, ideal circumstances without impurities can easily be modelled. Shen and Udell (1985) used a finite element method to calculate the liquid velocity for a given interface. After convergence the position of the interface was updated iteratively. Only for the initial iterations, underrelaxation of the interface update was necessary to obtain a stable solution. In the limit of Ca $\rightarrow$ 0, good agreement with Bretherton's lubrication approximation was found for both pressure drop and film thickness. A similar approach using finite difference discretisation of the governing equations was used by Reinelt (1987), again confirming Bretherton's theory. Irandoust and Andersson (1989b) used a finite difference approach in which the grid was rebuilt in a similar iterative approach, and incorporated Marangoni effects by using a no-slip boundary condition a certain distance from the flat film region (i.e., part of the transition region and the spherical cap region in Fig. 5.) and a no-shear boundary for the remainder of the interface. Edvinsson and Irandoust (1996) used a *false transient* FEM method using flexible grids, that again confirmed the Bretherton analysis for low *Ca*. The recent works of Giavedoni and Saita, 1997 and Giavedoni and Saita, 1999 and Heil (2001)have improved on the analysis of Edvinsson and Irandoust (1996) by coupling the translational velocity and the interface movement in each (steady state) iteration step. It seems that FEM methods involving flexible grids are the method of choice for analysing Taylor bubbles in capillaries. Volume-of-fluid methods are less often applied. For instance, Taha and Cui (2004) applied significant grid refinement and used very small timesteps in a transient simulation to obtain only reasonable results. Perhaps the reconstruction of the interface using VOF tends to introduce small errors in the curvature of the gas–liquid interface. In this surface tension dominated flow such errors lead to enormous pressure terms that destabilise the solution. Further, using the flexible grid methods one always has enough cells in transition region and film region, and the resolution with which the interface shape can be determined is many times higher than the number of cells per channel crosssection.

Concluding Remarks

The flow behaviour of Taylor flow is relatively well understood. The balance of the surface tension and viscous stresses provides analytical solutions to parameters of interest to chemical engineering applications, such as the pressure drop and the thickness of films. Deviations from the simple analysis due to Marangoni effects, gravity or inertia have been studied analytically or numerically, and where reliable experimental data is available, theory and simulations agree with experiment with a degree of accuracy rarely found in multiphase reactor engineering.

LIQUID FILM THICKNESS

Experimental data and fitted correlations are available for round channels, and to a lesser extent also for square channels. In monoliths in particular and coated microchannels in general, the corners are rounded to some extent by the coating process, and for rounded corners no film thickness correlations are available. Moreover, the rounding of the corners is sometimes not perfectly symmetrical, so defining a general film thickness correlations is going to be impossible unless we accurately describe the channel geometry. Here, we discuss the results for round and square channels, which gives a lower limit (round channels) and a upper limit (square channels).

Round Capillaries

For round capillaries, numerous experimental correlations of the film thickness are available (e.g. Chen, 1986, Irandoust and Andersson, 1989a, Thulasidas et al., 1997 and Aussilous and Quére, 2000). Most of these correlations tend to zero for vanishing Ca, and most exhibit a smaller slope on a graph of $\log(/r)$ vs. $\log(Ca)$, as in Fig. 6, than the lubrication value of 2/3. The change in slope can be attributed to the increasing importance of Marangoni effects at small Ca. Typically slopes between 0.5 and 0.66 are reported: Fairbrother and Stubbs (1935) found 0.5, Irandoust and Andersson (1989a) found 0.54, Halpern et al. (1998) found 0.52, etc. All these correlations give the same order of magnitude for the film thickness. The correlation of Aussilous and Quére (2000) for round capillaries,

$$\frac{\delta}{d} = \frac{0.66Ca^{2/3}}{(1 + 3.33Ca^{2/3})} \qquad (6)$$

has the benefit of accounting as best as possible for high Ca values and agrees with the data of Bretherton and Taylor for $10^{-4} < Ca < 10^{0}$.

Square Capillaries

For square capillaries, most correlations give comparable results for the film thickness in the corners. For the film far away from the corners, however, there is no such agreement. Kolb and Cerro (1991) measured the shape of the liquid film for different capillary numbers in the directions *AA'* and *BB'* (see Fig. 9). Note that the channel diameter is defined in the direction *BB'*, so in the direction *AA'* the maximum bubble size is s √2 times the channel diameter.

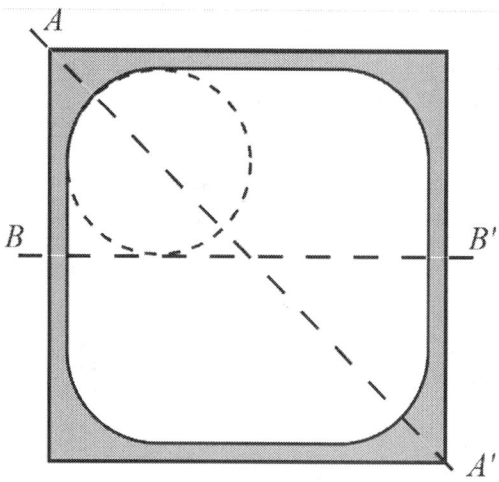

Figure 9: The shape of the liquid film in square capillaries for *Ca*<0.04 (Kreutzer, 2003).

Thulasidas et al. (1995a) measured the film thickness in the direction *AA* using optical methods for a wide range of capillary numbers. Hazel and Heil (2002) computed the shape of bubbles in square capillaries using a finite-element free surface formulation, similar to Heil (2001).

In Fig. 10, the experimental data (Kolb and Cerro, 1991 and Thulasidas et al., 1995a) and numerical data (Hazel and Heil, 2002) are plotted against the capillary number. The data presented in Fig. 10 are for horizontal flow. In square capillaries, the influence of gravity on the film thickness is more pronounced than in circular capillaries. For the diagonal (*AA*) direction, the agreement between all the data is good. For r Ca → 0, the dimensionless bubble diameter approaches 1.2. In other words, even at low velocities the film does not vanish completely in the corners. Note that the upper limit is based on the data of Thulasidas et al. alone. For Ca → ∞, the dimensionless bubble diameter approaches a value of 0.7. For r Ca < 0.04, the bubble diameter in both directions is the same and the bubble is axisymmetric, while the results in Fig. 10 show that for Ca < 0.04 the bubble diameter in the *BB* direction is virtually independent of Ca.

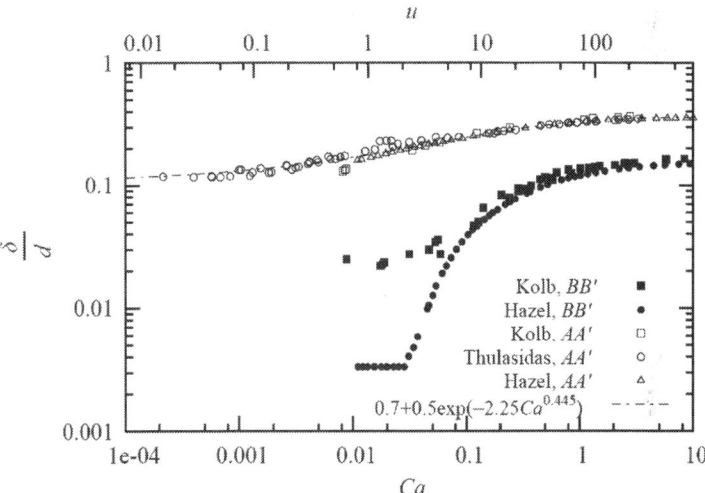

Figure 10: Film thickness versus *Ca* in square capillaries. Experimental data from Thulasidas et al. (1995a) and Kolb and Cerro (1991), numerical data from Hazel and Heil (2002). On the top axis, the velocity of the bubble is plotted, assuming water-like properties $_{\mu L}=10^{-3}$ Pa s and γ=0.073 N/m.

Using the asymptotic values of 1.2 and 0.7, the dimensionless bubble diameter in the diagonal direction can be correlated against the Capillary number as

$$\frac{d_{b,\text{square}}}{d_{\text{channel}}} = 0.7 + 0.5 \ \exp(-2.25 Ca^{0.445})$$

$$(7)$$

which is also plotted in Fig. 10. For monoliths, the region of interest is Ca < 0.04. Here the bubble diameter in the direction *BB* is close to the width of the channel. From the data of Hazel and Heil , a value of $d_{bubble}/d_{channel}$=0.99 is obtained, while the experimental data of Kolb and Cerro is somewhat lower, $d_{bubble}/d_{channel}$≈0.95.

BUBBLE VELOCITY

In Taylor flow, the bubbles travel slightly faster than the sum of the gas and liquid superficial velocity. If the bubble shape or the film thickness is known, the volumetric liquid flow rate can be calculated in the film. If needed, gravity can be incorporated to yield a falling film, see Grolman et al. (1996). Otherwise, the volumetric flow rate in the film is zero.

For round capillaries at low *Ca* the film is very thin. Using A_{bubble}→A and u_{film}→0,

$$A_{\text{bubble}} \to A \text{ and } u_{\text{film}} \to 0,$$

$$\frac{u_{\text{bubble}}}{u_{\text{slug}}} = 1 + \frac{4\delta}{d}.$$

$$(8)$$

In square capillaries, even at low *Ca* a finite film remains in the corners. The approximation of zero velocity in the film is therefore less appropriate, unless the channel is horizontal. In short, the exact value of the excess velocity is just as inaccurately known for coated channels as is the film thickness. On the other hand, if the film shape is known, the bubble velocity can easily be calculated.

STREAMLINE PATTERNS IN LIQUID SLUGS

The bubble is separated from the wall by a thin film that is also present between the slug and the wall. It is important to realise that the circulating vortex is never in direct contact with the wall, and mass (heat) can only be transferred to the wall by diffusion (conduction) through the film. To model these mass transfer steps, we need the slug film thickness.

Thulasidas et al. (1997) measured the location of the streamline dividing the circulating liquid in the slug from the film using PIV. The location of the dividing streamline is plotted in Fig. 11 together with the bubble diameter based on the low-inertia film thickness correlation of Aussilous and Quére (2000).

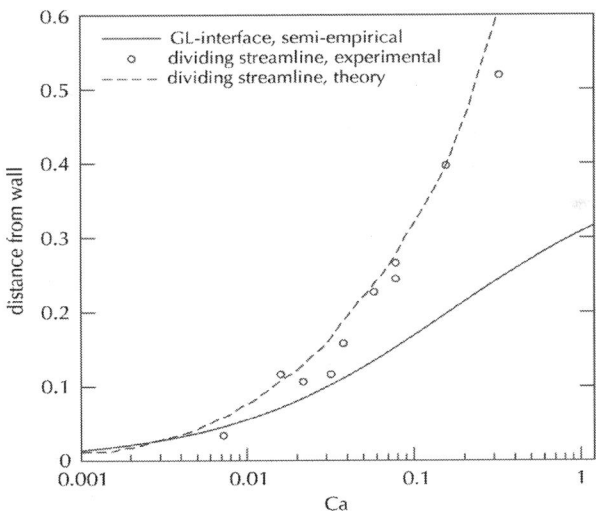

Figure 11: Comparison of the film thickness between the bubble and the wall with the film thickness between the slug and the wall. The solid line, representing the thickness of the film when a bubble passes by, is calculated from the low-inertia correlation of Aussilous and Quére (2000), which is based on experimental data. The markers are measurements of the location of the streamline dividing the film from the slug, and the

dashed line indicates the theoretical location of this streamline (Thulasi-das et al., 1997).

For circular capillaries at low *Ca*, the film thickness between the slug and the wall is comparable to the film thickness between the bubble and the wall. This is the region of interest for monoliths, and in a first approximation the bubble film thickness can be used for the slug as well. For Ca > 0.01, the film between the slug and the wall is thicker. For r Ca > 0.7, the circulating region completely vanishes.

In Fig. 12, calculated liquid streamlines are shown. The transition regions at the front and the rear of the bubble are enlarged, and the film region is shaded in gray. The white area is circulating: the bubble pushes the liquid ahead of itself. Within a tube diameter, the flow is practically developed into parallel Hagen–Poiseuille flow. Streamlines for very short slugs, up to (L/d)<0.25, have been reported by Fujioka and Grotberg (2004). In such short slugs, the average film thickness is higher, but still the circulation region is found.

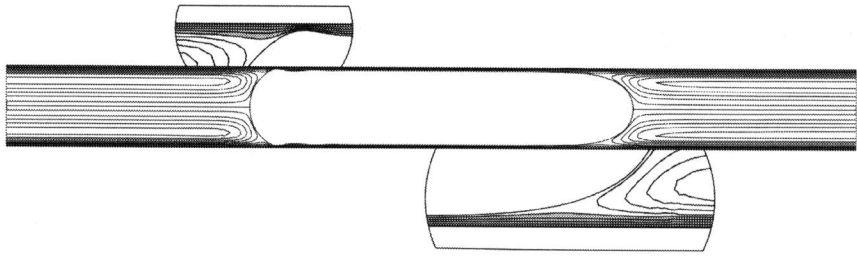

Figure 12: Streamlines in the liquid phase. The film region is indicated in grey and the circulating region is indicated in white (from Kreutzer, 2003).

PRESSURE DROP

Balancing Viscous Friction and Static Head

In Taylor flow, several contributions to the pressure drop must be considered. The first contribution is due to the wall friction of the slugs. In laminar flow this term is purely viscous, and the Hagen–Poiseuille value of

$$(\Delta p/L) = 32\mu_{uTP}/d^2 \qquad (9)$$

provides a reasonable estimate for the pressure drop per unit slug length. Here, the actual velocity of the slugs is estimated (assuming zero excess velocity) using the sum of the liquid superficial velocity and the gas superficial velocity,

$$u_{TP} = u_{Ls} + u_{Gs} \qquad (10)$$

To obtain the pressure drop per unit channel length, the contribution of the slugs to the pressure drop should be multiplied by $_{L}$, the fraction of the channel filled with slugs, estimated by

$$\varepsilon_L = u_{Ls}/u_{TP}. \qquad (11)$$

The second contribution is due to gravity. Per unit slug length this contribution is given by

$$(\Delta p/L) = \rho g. \qquad (12)$$

Again, multiplication with the holdup $_L$ gives the value per unit channel length. The third contribution is due to Laplace pressure terms, which can be estimated from lubrication theory using Eq. (5) for each bubble.

In the engineering literature, the first and second contributions are usually taken into account, leading to

$$\frac{\Delta p}{L} = \varepsilon_L \left[\frac{32\mu u_{TP}}{d^2} + \rho g \right] \qquad (13)$$

$$= \varepsilon_L \left[f_{\text{app}} \left(\frac{1}{2} \rho u_{TP}^2 \right) \frac{4}{d} + \rho g \right]$$

(14)

which is the same as the single phase equation used to define the friction factor f, except that (1) the sum of the gas and liquid superficial velocity, u_{TP}, is used to estimate the actual velocity of the bubble train as it passes through the channel and (2) the right hand side is multiplied by the fraction of the channel occupied by slugs, ε_L. If static head and wall friction are the most important contributions to the pressure drop, then the apparent friction factor f_{app}, as defined by Eq. (14), will be close to the Hagen–Poiseuille single-phase value of $16/Re$ for round channels.

The earliest attempt to model pressure drop in monoliths is by Satterfield and Özel (1977), who subtracted the static head from the total pressure drop to obtain a pressure drop that appeared to be dominated by wall friction. Grolman et al. (1996) using a model similar to Eq. (14). Their experimental data, reported as f_{app} defined by Eq. (14) from experimental data obtained in monoliths, showed that the pressure drop was consistently higher. The excess pressure drop was attributed to entrance losses. Heiszwolf et al. (2001a) found that ($\varepsilon_L f_{\text{app}}$) was between $18/Re$ and $22/Re$, and they attributed the excess pressure drop to secondary flow patterns, similar to entry-region flow.

The relative importance of gravity and friction can be estimated from the free fall velocity u_{ff}

$$u_{ff} = \frac{\rho g d^2}{32 \mu}$$

(15)

At u_{ff}, gravity and friction balance, i.e., the left hand side of Eq. (13) is zero.

Laplace Pressure Terms

A convenient property of a simple balance of viscous wall friction and static head is that both are proportional to the holdup ε_L. The third important contribution, the Laplace pressure term, is proportional to the number of bubbles per unit length. This parameter is hardly ever reported alongside pressure drop data.

Fig. 13 shows the dimensionless pressure on the wall from a numerical study (Kreutzer et al., 2005b). The zero slope of the wall pressure in the bubble region indicates that in this region all pressure losses may be ignored. Further, the slope of the pressure in the slug regions in Fig. 13 indicates that Hagen–Poiseuille holds, except close to the bubble. Finally, the Laplace terms due to the presence of the bubbles are indicated in Fig. 13 by ξ.

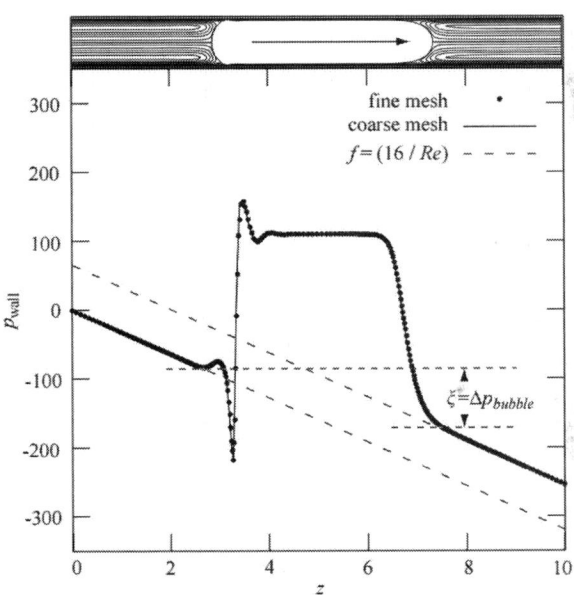

Figure. 13: Wall pressure in the axial direction for Ca = 0.01, Re = 100, L = 0.65, L* = 10. The axial coordinate is scaled with the channel diameter, and the pressure scale is viscous, i.e., the Hagen–Poiseuille frictional pressure drop may be represented by (p/z) = −32 (from Kreutzer et al., 2005b).

The lubrication analysis of Eq. (5) can be rewritten on a viscous pressure scale to obtain an apparent friction factor

$$f_{app} = \frac{16}{Re}\left[1 + \frac{d}{L_{slug}}\frac{0.465}{Ca^{1/3}}\right] \quad (16)$$

with f_{app} as defined by Eq. (14) per unit slug length. Note that the Laplace term is now lumped into the slug friction factor, and the relative importance of both contributions can be estimated by comparing 1 to $(0.465d)/(Ca^{1/3}L_{slug})$. For typical flows in monoliths and microreactors, $Ca^{1/3} = O(10^{-1})$ and—purely based on a scaling analysis using lubrication theory—the Laplace terms are significant if $L_{slug}/d < O(10^1)$,

Horvath et al. (1973) measured the pressure drop at different slug lengths, *ceteris paribus,* and found that the pressure drop increased when the slugs where shorter than 15 times the channel diameter. A plot of Δp versus (L_{slug}/d) was consistent with Eq. (16) in the sense that it showed, for constant ε_L and u_{TP}, that $\Delta p \sim a + b/L_{slug}$. Horvath et al. discussed the slug length dependence in relation to mass transfer from short slugs, and were unaware of the Laplace terms. Fujioka and Grotberg (2004) calculated numerically the pressure drop across a slug and found that for short slugs, the Laplace terms dominate, and hence that the pressure drop is slug-length dependent.

Kreutzer et al. (2005b) performed measurements similar to Horvath et al., varying bubble and slug length independently. By using different liquids, *Re* and *Ca* could also be varied independently. The experimental pressure drop data, $\Delta p/L$, were rewritten as a friction factor based on the liquid-filled part of the channel, i.e., as f_{app} defined by Eq. (14). The group $f_{app}Re$ was independent of the bubble length, or—in other words—the only contribution of the bubble to the pressure drop occurs at the caps (ξ in Fig. 13) and is unrelated to the length of the bubble.

In contrast, $f_{app}Re$ strongly depends on the slug length, as predicted from lubrication. Fig. 14 shows how $f_{app}Re$ depends on the slug length for several liquids. The data show that for long slugs,

$$\lim_{L_{slug}\to\infty} f_{app} = 16/Re \quad (17)$$

so the single phase asymptote is obtained, in accordance with Eq. (16). More importantly, it shows that in (surface tension dominated) Taylor flow the Laplace pressure terms easily double the apparent friction factor for slugs shorter than 10 tube diameters. Rather surprisingly, Fig. 14 also shows that the apparent friction factor is a function of fluid properties and slug length only and apparently is independent of velocities. Kreutzer et al. (2005b) showed that inertia can be taken into account with the dimensionless group

$Re/Ca = \mu^2 / \rho d\gamma$, which is independent of velocities, and all data could be accurately represented by

$$f_{\text{app}} = \frac{16}{Re}\left[1 + 0.17\frac{d}{L_{\text{slug}}}\left(\frac{Re}{Ca}\right)^{1/3}\right]$$

(18)

This equation holds for $We > 0.1$. CFD simulations confirmed the use of Re/Ca to account for inertia, except for the constant 0.17. The constant 0.17 in Eq. (18) was obtained experimentally and includes the maximum increase (a factor $4^{2/3}$) in per-bubble pressure drop due to Marangoni effect. Without such impurities, 0.07 should be used in Eq. (18).

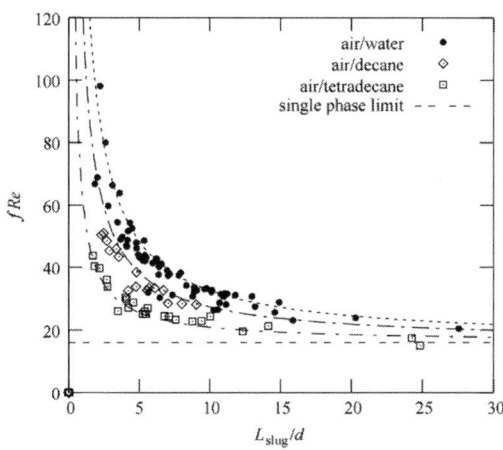

Figure 14: (fRe) as a function of dimensionless slug length. Channel diameter $d_c = 2.3$mm [from Kreutzer et al. (2005b)].

Concluding Remarks

Clearly, the pressure drop in Taylor flow is slug-length dependent. This slug length dependence may also be written in terms of the bubble frequency (Ratulowski and Chang, 1990 and Cubaud and Ho, 2004): more bubbles per meter channel lead to a higher pressure drop. The wide variety in slug lengths in different systems explains the wide range of experimental pressure drop values.

The slug length in channels is for many systems hard to predict a priori, but is typically between two and ten times the channel diameter. This makes it difficult to know the pressure drop in advance. On the other hand, experimental pressure drop data can be used in opaque systems where it is difficult to determine the slug length otherwise. Of course, if the channel inlet is constructed in such a way that the slug length can be controlled, correlations such as Eq. (18)(Re>100) and Eq. (16)(Re<10) are very accurate.

MASS TRANSFER

In Fig 15, three different mass transfer steps for a gas-component to the catalyst can be identified: (1)$_{kGSaGS}$, the transfer from the bubble through the liquid film directly to the catalyst, (2) $_{kGLaGL}$, the transfer from the caps of the gas bubble to the liquid slug and (3) $_{kLSaLS}$, the transfer of dissolved gas from the liquid slug to the catalyst. For a liquid component, only the last step (3) needs to be considered. In the absence of a catalyst on the channel wall, steps (1) and (2) both contribute to the physical absorption of gas.

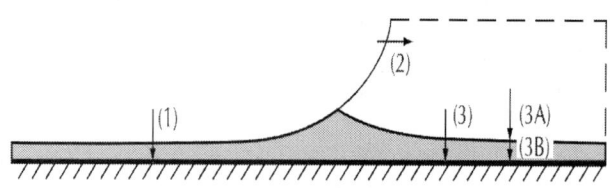

Figure 15: Different mass transfer steps in Taylor flow. (1) From the bubble

directly to the wall, (2) from the bubble to the vortex region in the slug, (3) from the vortex region to the wall. Note that the third step may be decomposed into a convective–diffusive contribution in the vortex region and a pure diffusive contribution in the film region.

In the simplest approximation for gas-to-catalyst mass transfer, we ignore all possible overlap and interaction between these transfer steps. The last two steps can then be considered as resistances in series and are in parallel with respect to the first step, and for the overall mass transfer we have

$$k_{GLS}a_{GLS} = k_{GS}a_{GS} + \left(\frac{1}{k_{GL}a_{GL}} + \frac{1}{k_{LS}a_{LS}}\right)^{-1} \quad (19)$$

In Taylor flow, all film regions do overlap and interact, and the concept of a liquid phase between the gas and solid does not hold. Therefore, one cannot independently measure gas–liquid mass transfer and liquid–solid mass transfer and then combine the results using (Eq. 19).

Because of the regular and well defined geometry of Taylor flow, computational fluid mechanics can be applied with relative ease: instead of having to calculate the gas–liquid interface, it can be estimated with correlations for film thickness from the previous sections. Then, the complex multiphase problem reduces to a laminar, single phase problem which can be readily solved with standard codes.

In this section, first the mass transfer in capillaries without reaction from gas to liquid is considered. Subsequently, mass transfer with reaction on the wall is considered, first for a liquid component and then for a gas component.

Physical Absorption of Gas

Gas components can be transferred to the flat film region and to the liquid slugs directly. It is important to realise that these two liquid zones do not mix, and transfer of matter between them occurs only by diffusion from the film to the slug.

Berčič and Pintar (1997) measured gas–liquid mass transfer in a single channel for a wide range of superficial gas and liquid velocity. Their experimental set-up allowed the independent variation of bubble and slug length. They correlated their data for a methane–water system as

$$k_L a = \frac{0.111 u_{TP}^{1.19}}{\left[(1 - \varepsilon_G)(L_{\text{bubble}} + L_{\text{slug}})\right]^{0.57}}$$

(20)

Note that in the denominator, $(1-_{\varepsilon G})$ $(_{\text{Lbubble}}+_{\text{Lslug}})$ is practically equal to the slug length. Interestingly, the mass transfer from Eq. (20) is a function of the slug length and hardly a function of the bubble length. Higbie (1935) used penetration theory for experimental mass transfer from a single Taylor bubble in capillaries, in which the contact time was estimated from the bubble length. A possible explanation for the completely different behaviour of bubble-train Taylor flow may be offered by assuming that the lubricating film near the wall is completely saturated each time the bubble passes by. Then, a longer bubble does not improve mass transfer and the mass transfer becomes independent of bubble length. The data of Berčič and Pintar were obtained for very long bubbles and slugs for which complete saturation of the film was likely. Then, the data of Berčič and Pintar describe (1) the partial depletion of the film between the wall and the slug as the slug passes by and (2) the transfer of gas to the slug at the bubble caps. The specific interfacial area associated with transfer from the caps is independent of channel diameter. Berčič and Pintar varied the channel diameter between 1.5 and 3.1 mm and found no impact of channel diameter. In monolith reactors, Kreutzer et al. (2001) found no difference in mass transfer between monoliths with small and large channels, which is in agreement with Eq. (20), and which suggests that transfer from the caps is important.

Irandoust et al. (1992) modelled gas absorption in Taylor flow. They assumed a penetration theory for the film between the bubble and the wall, and found agreement with experiment with limited adjustable curvefitting parameters.

van Baten and Krishna (2004) performed a CFD study of gas absorption in Taylor flow, and found that in some of the experiments of Berčič and Pintar (1997), the contact time in the film was long enough to fully saturate the liquid film. For shorter unit cells (or higher velocities), they formulated a mass transfer model of penetration theory for both the caps and the film

$$(k_L a)_{cap} = \frac{8\sqrt{2}}{\pi L_{UC}} \sqrt{\frac{D u_B}{d}}$$

(21)

$$k_{L,film} = \begin{cases} 2\sqrt{\dfrac{D}{\pi t_{film}}} \dfrac{\ln(1/\Delta)}{1 - \Delta} & : \ Fo < 0.1, \\ 3.41\dfrac{D}{\delta} & : \ Fo > 1, \end{cases}$$

(22)

$$a_{film} = \frac{4 L_{film}}{d L_{UC}},$$

(23)

in which the Fourier number Fo and the parameter Δ are defined by

$$Fo = \frac{D t_{film}}{\delta^2} \quad \text{and estimate} \quad t_{film} \approx \frac{L_{film}}{u_B},$$

(24)

$$\Delta = 0.7857 \exp(-5.212 Fo) + 0.1001 \exp(-39.21 Fo) + \cdots .$$

(25)

Note that for short contact time, the mass transfer group now becomes a function of the channel diameter. In the majority of the simulations performed by van Baten and Krishna (2004) the slugs were significantly longer than the bubbles, so depletion of the film in the slug region is likely. For gas absorption without reaction (at the wall or in the liquid), the alternating exposure of the lubricating film to bubbles and slugs periodically fills and empties this film, and the relative lengths of the bubbles and slugs determines which has the most impact. This explains why different engineering correlations are found, some based on slug length, but others based on bubble length: the experimental range of bubble and slug lengths determines which correlation best fits the data, and extrapolation of such correlations beyond the experimental bubble and slug contact times must be done with caution.

Liquid-To-Wall Mass Transfer

Now consider the transfer of a liquid phase component to a catalyst on the wall. The best approach would be to consider two different mass transfer steps, one from the circulating vortex to the film, in series with a second resistance inside the film. The first step can be considered by eliminating the film from consideration in a numerical study, while experimentally the film resistance can be eliminated by choosing the capillary number sufficiently low to have a negligible film thickness.

The principal features of the first mass transfer step can then be studied by ignoring the thin film, and simplifying the gas–liquid interface to flat ends. Duda and Vrentas (1971b) used this approach and found an infinite-series analytical solution for the closed-streamline axisymmetric flow in this cylinder. In a second paper (Duda and Vrentas, 1971a), the corresponding developing heat transfer problem was solved using a formal Fourier series technique. The method allowed the calculation of time dependent Nusselt numbers up to $(L_{slug}/d)=2.5$ for Peclet numbers of up to 400. Extension to higher (L/d) was prohibited as the eigenvalues of the solution became too close together as the aspect ratio was increased.

Analogous to the single phase Graetz problem, the Graetz number $Gz = L_{ch}/d Re\, Sc$ can be introduced. Kreutzer (2003) calculated the liquid–solid mass transfer in this simplified geometry with a finite-element method, arriving at different values than reported by Duda and Vrentas (1971a). The results of Kreutzer (2003) gave an expression for the length-averaged mass transfer from a circulating vortex to the wall, without a lubricating film in between:

$$Sh = \sqrt{\alpha^2 + \frac{\beta}{Gz}}$$

(26a)

with α and β weak functions of the slug length

$\alpha = 40(1 + 0.28(_{Lslug}/d)^{-4/3})$, (26b)

$\beta = 90 + 104(_{Lslug}/d)^{-4/3}$, (26c)

Eq. (26a) is defined per unit slug volume and should be multiplied with the liquid holdup to obtain a mass transfer coefficient based on channel volume. Also, Eq. (26a) is only valid for the region in which the circulating vortex has at least circulated once. Before a full circulation the effect of circulation has hardly manifested itself, and the Sherwood numbers for very short tubes are lower and they are complex functions of slug length and tube length.

Closer inspection of Eq. (26a) shows that even for long slugs, the asymptotic value for $G_z \rightarrow \infty$ is 40, which may be compared to 3.66 for the analogous single phase case. So, by adding bubbles we can increase our liquid–solid mass transfer by a factor 10 (provided the film does not become limiting) per unit liquid volume. With a holdup of 50%, still a factor 5 remains. The Sherwood number may quite generally be interpreted as a dimensionless gradient at a boundary. In single phase flow, the distance between the minimum and maximum concentration is equal to a channel radius, while in Taylor flow the largest difference is found between the wall and the vortex centre, located $r\sqrt{2}$ from the axis, i.e., over a distance of 0.293r. Kreutzer (2003) attributed the enhanced mass transfer in part to this shorter distance, and in part to the fact that circulation prevents liquid at the core of the channel to flow out of the channel at high velocity, in other words, that all fluid elements are convected to the channel wall and have a similar residence time. Gruber and Melin (2003) performed a numerical study of liquid–solid mass transfer in Taylor flow, and experimentally studied mass transfer by dissolving a copper capillary in a sulphuric acid/potassium dichromate solution. Gruber and Melin (2003) considered the entire unit cell in their analysis, and found that the film resistance could be ignored if $_{\text{dfilm}}/d<0.01$. In a CFD study, van Baten and Krishna (2005) used $_{\text{dfilm}}/d=0.0016$ for simulations comprising a complete unit cell, including the bubble. They based Sh on the entire unit cell surface area and modified the definition of Gz to Gz to $Gz = [(1 - G) L_{\text{ch}}/d\text{Re Sc})$. Their results could be correlated as

$$ Sh = 0.5 \left(\frac{\varepsilon G}{Gz} \right)^{0.15} Gz^{-\alpha}, $$

(27)

where ≈ 0.48 is a weak function of slug length. The results of this study agreed with the experimental data of Horvath et al. (1973) (see below). Note that Eq. (27) does not contain the film thickness as a parameter. Locally, the isolating effect of the lubricating film was observed, but for the range of operating parameters used by van Baten and Krishna (2005), the effect on the overall (unit-cell averaged) mass transfer was limited. If the slugs and bubbles were very long (~ 50 mm) compared to the channel diameter of 1.5 mm, the fluxes were significantly higher close to the slugs.

Limited experimental data for liquid–solid mass transfer are available. Oliver and Youngh-Hoon (1968) measured heat transfer in two-phase flow in capillaries using very viscous liquids and hence with thick lubricating layers. Hatziantoniou and Andersson (1982) and Ber i and Pintar (1997) measured the dissolution of benzoic acid in short tubes, where the liquid had not always circulated at least once.

Horvath et al. (1973) measured the hydrolysis of N-benzoyl-arginine ethyl-ester in a 1.2 m tube coated with the immobilised enzyme Trypsin. The intrinsic rate of this reaction was high enough to ensure that mass transfer from the liquid was limiting. The experimental data are reported as Sh versus $(_{\text{Lslug}}/d)$ with Re as a parameter and Sh versus Re with the aspect ratio $(_{\text{Lslug}}/d)$ as a parameter. The Sherwood numbers were corrected for the fact that they were calculated based on the total area in the tube, i.e., including the parts of the tube that were occupied by gas bubbles. In Fig. 16, the experimental data are compared with the results of Kreutzer (2003). The agreement of the cylindrical cavity calculations is very good for low Reynolds numbers. Note that at low Reynolds numbers, the capillary number is also low. For higher *Re*, the impact of the film resistance increases.

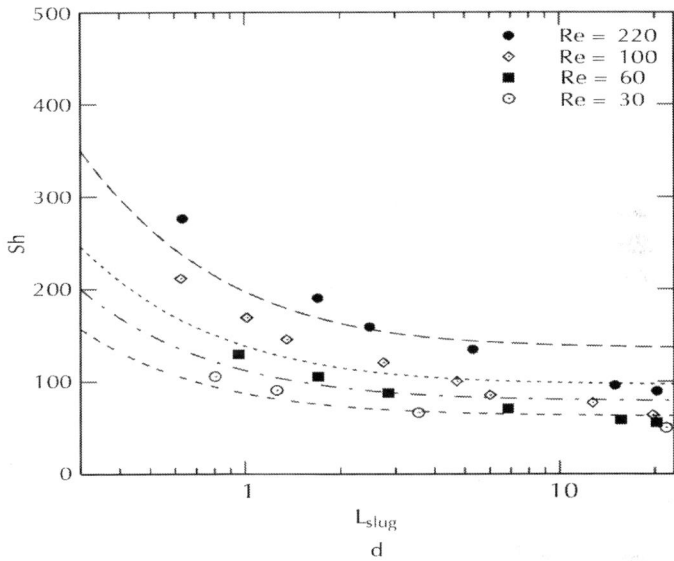

Figure 16: Sherwood number plotted against the slug aspect ratio. Experimental data from Horvath et al. (1973), lines based on Eq. (26a). Note that the applicability of Eq. (26a) is based on a negligible film resistance, which is satisfied here only for low *Re* (from Kreutzer, 2003).

Mass Transfer under Reacting Conditions

Perhaps because of the intended comparison with trickle–bed reactors, the first experimental studies of three-phase reactions in monoliths were performed at such low superficial velocities (e.g. Mazzarino and Baldi, 1987), that the falling film flow pattern was more likely than Taylor flow.

Smits et al. (1996) hydrogenated 1-octene and styrene in a 400 cpsi monolith reactor with washcoated square channels. The superficial velocities were in the range $0.04 < u_{TP} < 0.45$ m/s. The observed reaction rate first increased with increasing total velocity. Smits et al. interpreted this as an improvement of the distribution with increasing flow rate. When the linear velocity u_{TP} was increased further, the observed reaction rate dropped. Nijhuis et al. (2001b) hydrogenated α-methylstyrene and benzaldehyde over monolithic Ni catalyst in a pilot-scale reactor. The experimental finding of Hatziantoniou et al. (1986) that the holdup has little impact on the observed reaction rate was confirmed.

Broekhuis et al. (2001) conducted an unspecified nitro-aromatic hydrogenation in an internal circulation (Berty) autoclave, and measured a pseudo-first order rate constant for hydrogen of 2.5 s^{-1}. Kreutzer et al. (2001) hydrogenated α-methylstyrene over a Pd catalyst. The pseudo first order rate constant for hydrogen was well above 1 s^{-1}.

Kreutzer (2003) used a model based on Eq. (19) to estimate that the dominant resistance for hydrogen transfer to the catalyst was in the lubricating film. This film resistance, for which a mass transfer coefficient may be estimated from film theory, i.e., $k=D/\delta$ for film thickness δ, is found between the bubble and the wall and between the slug and the wall. Kreutzer et al. found no impact of holdup experimentally. At lowCa, the lubricating film thickness is comparable for the slug and for the bubble (see Figs. 11 and the enlarged parts of 12 and, for a more in depth discussion, Thulasidas et al. (1997)). Below $Ca = 0.01$, the difference in thickness vanishes, and then, the transfer of hydrogen through the film is about as fast for the slug as for the bubble, i.e.,

$$k_{\text{bubble-wall}} = \frac{D}{\delta_{\text{bubble-wall}}} \approx \frac{D}{\delta_{\text{slug-wall}}} = k_{\text{slug-wall}},$$

(28)

while for higher Ca the data of Fig. 11 might be used to correct for the thicker film for the slug.

Fig. 17 shows that the largest resistance to mass transfer in the slug is located in the thin film region, and that the majority of the circulation zone is characterised by a region of constant concentration. For the bubble region, the mass transfer can be modelled using film theory

$$J_{GS} = kA(c^* - 0) = \frac{D}{\delta} \frac{4L_{\text{bubble}}}{d}(c^* - 0).$$
(29)

For the slug region, the same approach could be used, with the concentration in the circulating zone

$$J_{LS} = kA(c_{\text{slug}} - 0) = \frac{D}{\delta} \frac{4L_{\text{slug}}}{d}(c_{\text{slug}} - 0).$$
(30)

So, the problem of formulating a mass transfer model was reduced to the problem of formulating a model that predicts the average slug concentration. For short slugs and typical hydrogenation conditions of organic liquids in square channels, the slug is almost saturated and $c_{\text{slug}} \approx c^*$. For very long slugs, this needs not to be the case.

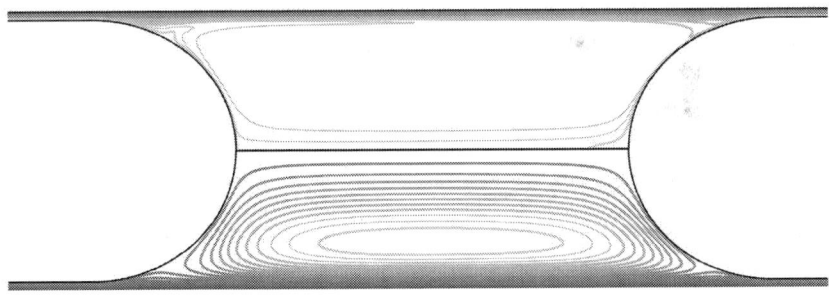

Figure 17: Concentration contours of 20 equally spaced intervals between 0 and c^* (top half) and streamlines (bottom half) for a simulation with D=1.4×10⁻⁸ m²/s, u_{TP} =0.4 m/s, , dchannel = 1.0 mm, $L_{\text{slug}} + L_{\text{bubble}} = 4 d_{\text{channel}}$ and holdup $_L$=0.5 (fromKreutzer, 2003).

An important finding that is consistently reported in the literature is explained by the CFD simulations: the impact of holdup is limited. If the conditions are such that the slug is completely saturated with the gas–phase component, and if the film thickness is the same for the slug and the bubble, the mass transfer is indeed completely independent of the holdup.

Fig. 18 shows another striking aspect of mass transfer to the wall in capillary or monolith columns: with decreasing velocity the external mass transfer is improved. This implies that the mass transfer improves with a decrease in pressure drop. This behaviour is of course related to the film thinning at lower Ca, but it is also radically different from intuitively expected behaviour. The notion that enhancement of mass transfer comes at the cost of an increase in pressure drop is almost an axiom in reactor engineering, formalised for instance by Colburn (1933). It should be realised that such analogies are based on the dominance of eddy transport in turbulent flows, and the behaviour of the microchannel is in no means in contradiction with such analogies. The excellent mass transfer at minimal power input is one of the nicest features of monoliths in particular and multiphase microchannel in general, allowing an escape from the all too common trade-off of pressure drop and mass transfer.

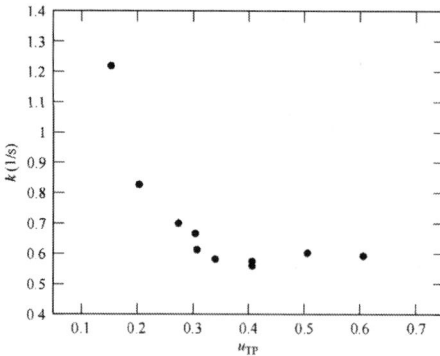

Figure 18: Observed pseudo-first order reaction rate constant for the hydrogenation of α-methylstyrene in a monolith pilot reactor.

The reaction was not completely mass transfer limited, but external mass transfer limitation did strongly affect the observed rate: for these experiments, $_{kobs,H2} \approx 0.5 _{kGLS}$. Note that the reaction rate decreases with decreasing throughput. (reprinted fromKreutzer et al., 2001, copyright (2000), with permission from Elsevier).

RESIDENCE TIME DISTRIBUTION

In Taylor flow, the degree of backmixing is diminished by the presence of the bubbles, which effectively seal packets of liquid between them. In fact, the only mechanism for transfer of matter from one slug to the next is by diffusion from the slug to the film, and subsequent diffusion from the film to the next slug. The combination of enhanced radial mass transfer (local mixing) and suppressed backmixing makes Taylor flow an ideal hydrodynamic regime, not only for (gas–) liquid–solid reactions, but also for homogeneous reactions where plug flow behaviour is essential (see e.g. the use of Taylor flow in the synthesis of colloidal silica by Khan et al. (2004)).

Thiers et al. (1971) studied the inter-slug mass transfer in a capillary. In their experiments, the slugs were so long that complete equilibrium between the slug and the film could be assumed, and a simple tanks-in-series model described their data. For shorter slugs, the complete mixing of bubble and slug cannot be assumed. To a first approximation, the lubricating film may then be regarded as stagnant, and the isolated slugs may be considered in plug flow. Matter exchanges between these two zones, which leads to the classical piston-exchange model that was first introduced by Hoogendoorn and Lips (1965).

The piston-exchange model of residence time distribution leads to *E*-curves that are significantly different from axial-dispersion curves. Axial dispersion describes stochastic, random fluctuations of velocities around a mean value, where higher velocities are just as likely as lower velocities. The resulting *E*-curve is then symmetric around the peak, i.e., the rise is just as long or short as the tail. In the two-zone piston-exchange description, the exit of tracer from the system can only be delayed by exchange with the "dead" stagnant zone, and never can this exchange result in an earlier exit. The corresponding *E*-curve for Taylor flow, therefore, has a relatively long tail and a sharp rise, where the tail describes the leaking to upstream slugs. Fig. 19 shows an experimental curve from Thulasidas et al. (1999) that has these features. In fact, Thulasidas et al. reported only the tail of the curve; the first reported experimental concentration was the highest. Presumably, no tracer left the system earlier than the slug that was injected with tracer.

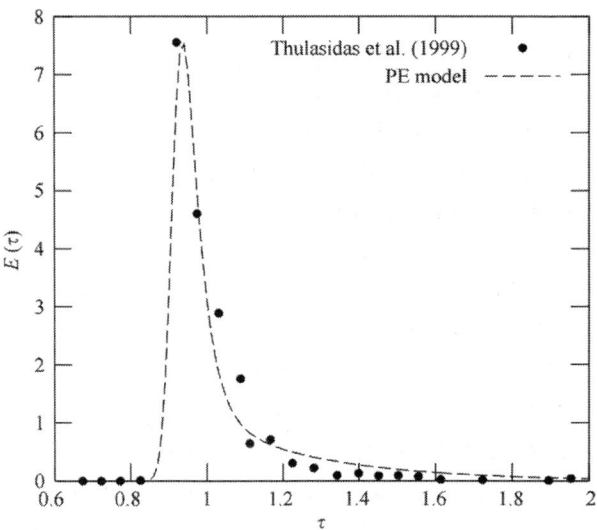

Figure 19: Example RTD curve, from Thulasidas et al. (1999). The symbols indicate the experimental data. The line is calculated using the PE-model of Hoogendoorn et al., using the high-*Gz* limit of Eq. (26a) to calculate the mass transfer rate and Eq. (6) for film thickness (i.e., the static fraction) (from Kreutzer, 2003).

Several similar two-zone models have been put forward, that are different only in the estimate of the exchange rate between the two zones. Pedersen and Horvath (1981) used the liquid-wall mass-transfer data of Horvath et al. (1973) to estimate the exchange rate, and then successfully used a two-zone piston-exchange model to describe back-mixing in Taylor flow. Thulasidas et al., 1995b and

Thulasidas et al., 1999 improved the analysis by modelling the lubricating film as a falling film. In up-flow in square channels, with the film between bubble and wall flowing downwards, the convective contribution of this falling film to the overall backmixing was significant. An infinite series solution was used for radial diffusion and improved the mathematical model further.

In straight channels, such as those found in monoliths, the piston-exchange behaviour leads to the very long tail that is so characteristic of dead zones. In meandering channels, which are more common in microreactors, a much more symmetric residence time distribution was found by Günther et al. (2004), i.e., the front and tail of the E-curve are of similar width. Khan et al. (2004) observed that the merging of slugs in the sharp 180° corners of a serpentine channel layout broadened the E-curve, which is a stochastic phenomenon that leads to a symmetric curve. Then, the strong tailing in of straight channels indicates the absence of slug merging in those straight channels.

Concluding Remarks

There is considerable agreement on how to describe RTD in Taylor flow, and the experimental data agrees reasonably well with a model calculation based on order-of-magnitude estimates for the static fraction and the film-slug exchange rate (Fig. 19). As improved models for the exchange rate become available, the RTD model will improve accordingly.

The application of the study by Thiers et al. (1971), separation of samples in medical analysis, indicates that carry-over of material from one slug to the next is minimal. For engineering purposes, the E-curves for single capillary channels may be regarded as plug flow for all but the highest conversion levels.

For the application in larger systems, the distribution of exit age that results from differences between the channels, as opposed to within a single channel, must also be accounted for and this contribution will be discussed next. Arguably, the differences between channels will be more symmetric, and the very sharp rise of the single-channel E-curve will be smeared out in systems with parallel channels. In other words, the combined study of both the single channel behaviour and the array of parallel channels provides clues to the extent of maldistribution.

SCALE UP OF CAPILLARIES TO MONOLITHS

So far we have only discussed the hydrodynamics and transport phenomena in a single channel. When we use a monolith block, i.e., an array of parallel capillaries channels, the question presents itself whether all these channels will essentially behave in the same way. First of all, the channels themselves can be slightly different, and slight variations in channel diameter will result in slight variations in fluid velocity (Valdés-Solís et al., 2004 and van Gulijk et al., 2005). Second, the distribution of gas and liquid into the channels can vary from channel to channel. Distribution issues are known for most multiphase reactors, but in monoliths the problem is aggravated by the fact that there is no convective transport from one channel to the next. In other words, an imperfect distribution just propagates downwards, and is not "corrected" or levelled-out in the column. Of course, the opposite is also true: a good initial distribution is not spoilt while propagating down the reactor.

Hydrodynamic Stability

Grolman et al. (1996) modelled the hydrodynamic behaviour of a large assembly of channels by requiring that in all channels, the pressure drop is the same. Perfect gas–liquid distribution results in all channels operating at identical u_{Ls} and u_{Gs}. With maldistribution, some channels receive more liquid than others, but all channels must operate on the same pressure gradient curve. In general, hydrodynamic stability is obtained if the response to an increase in flow rate is an increase in the resistance to flow, or

$$\frac{\partial}{\partial u}\left(\frac{\partial p}{\partial L}\right) > 0.$$

(31)

So, accurate knowledge of the pressure drop is required to determine the stable operating conditions. Kreutzer et al. (2005a) used the same analysis as Grolman, but included the Laplace terms in the pressure drop model. As a result, the unstable area on a map of u_{Ls} vs. u_{Gs} was a order of magnitude smaller than based on the balance of viscous friction and gravity. Kreutzer et al. (2005a) found that up-flow is always unstable: a slight increase of gas flow into a channel decreases the liquid hold-up in that channel. This lower hold-up creates a slight underpressure at the bottom entrance of the channel, which in turn draws even more bubbles into the channel. Downflow becomes unstable at low flow rates by a similar argument. The most important parameter for the stability analysis was the free fall velocity u_{ff} (see the section on pressure drop earlier). When u_{TP} is significantly lower than u_{ff}, gravity dominates and instabilities occur. When u_{TP} is on the same order as u_{ff}, the viscous (and Laplace) contributions to the pressure drop, for which Eq. (31) always holds, are dominant enough to ensure stable operation. Because the pressure drop is the key to stable operation, the bubble frequency is an important parameter for the stability, and distributors for downflow monoliths should be designed such that small bubbles are generated.

Residence Time Distribution

The residence time of a liquid tracer is primarily determined by u_{TP}, which gives more or less the velocity of the liquid slugs in laboratory coordinates. Up-flow RTD experiments have been performed by Thulasidas et al. (1999). The results of a single capillary were compared to the results for a bundle of square capillaries simulating a monolith. The results showed that in up-flow, the monolith was almost completely back-mixed, and the single channel data did not agree at all with the monolith data. Fig. 20 shows that the laminar parabolic flow of liquid alone gave an even better RTD than Taylor flow. Recently, Gladden and co-workers (Gladden et al., 2003, Gladden, 2004 and Mantle et al., 2002) have visualised up-flow in monoliths at low superficial velocities using MRI tomography. In most of the channels, the flow was indeed upward, but a wide range of velocities was found. Further, in some channels, the direction of flow was downward, resulting in recirculation over the monolith block. This recirculation behaviour, combined with the spread in velocities, explains the large extent of back-mixing observed by Thulasidas et al. (1999).

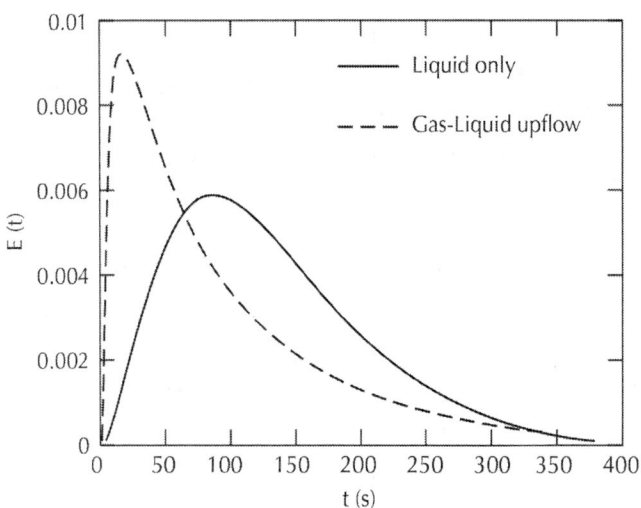

Figure 20: Experimental normalised concentration distributions for only-liquid flow and Taylor flow in a capillary bundle (reprinted fromThulasi-das et al., 1999, copyright (1999), with permission from Elsevier).

For downflow at higher velocities, the situation is different. An increase in liquid velocity is accompanied by a decrease in gas velocity, and as a result the distribution of u_{TP} , and thereby the RTD, is less broad than the distribution of u_{Ls} and u_{Gs} . Kreutzer et al. (2005a) measured the RTD of monolith columns in downflow (Fig. 21), and found that in downflow the deviation from single channel behaviour is not too large. By regression of the data to a PDE model, which may be interpreted as a two-zone single channel model with a Gaussian maldistribution superimposed, it was found that the contributions to the RTD of the two-zone single channel behaviour and the maldistribution were comparable. Also, experiments with multiple monolith blocks of different length showed that stacking of monolith blocks diminishes deviation from plug flow: between monolith blocks redistribution occurs. Later, Yawalkar et al. (2005) varied the channel diameter in monolith RTD experiments, and their experimental data suggested that tailing was reduced for smaller channels. Using the piston-exchange model, Hoogendoorn and Lips (1965) showed that the variance of a RTD curve due to exchange with a stagnant zone is inversely proportional to the mass transfer group (k·a) that describes the exchange between the dynamic and the stagnant zone. Naturally, a increases with decreasing channel size, so the reduced variance for smaller channels is consistent with the two-zone exchange model that was used successfully for the single-channel curves.

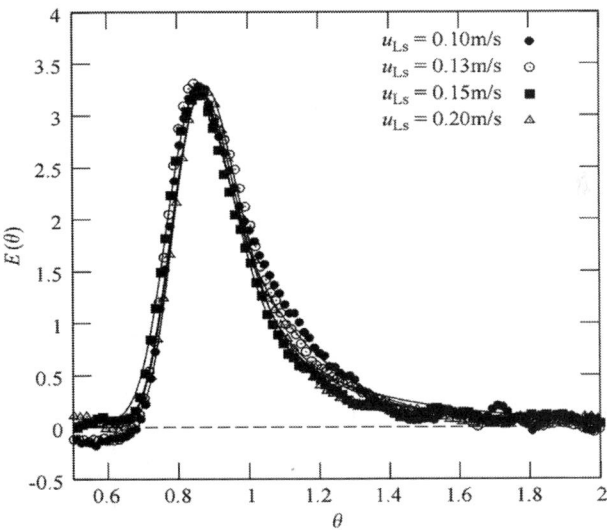

Figure 21: Experimental E -curves for downflow in a monolith column. The liquid superficial velocity $_{uLs}$ is varied at u Gs ≈ 0.15 m/s (from Kreutzer et al., 2005a).

Slug Length and Distributors

Kreutzer et al. (2005c) showed that their pressure drop model could be used to estimate the slug length monoliths based on pressure alone. This method was favourably compared to independent measurements of the slug lengths in the same setup by conductivity (Heiszwolf et al., 2001b).

Fig. 22 shows an experimental pressure map for a 200 cpsi monolith. The shaded area is the unstable region. Since mass transfer improves with decreasing velocity, the most optimal design of a downflow monolith reactor is often such that the flow is just stable, and an accurate estimate of the bubble length is essential if one would like to operate close to this limit.

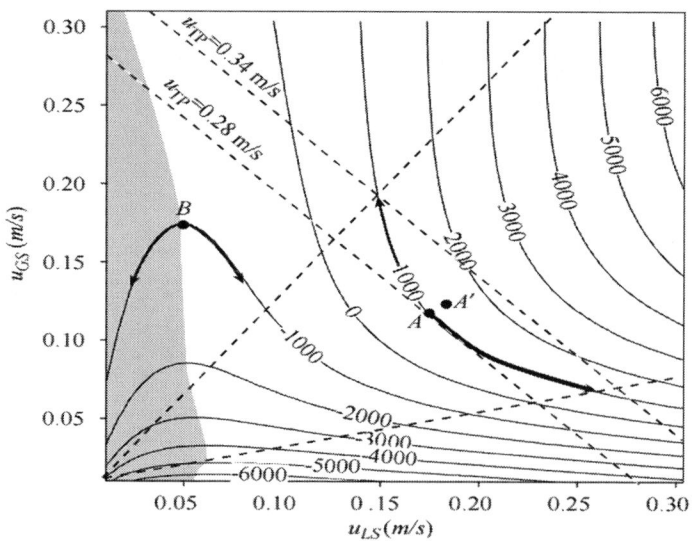

Figure 22: Solid, operating lines of constant pressure gradient (Pa/m) and dashed lines of constant liquid hold-up $_{\varepsilon L}$ in the $_{uLs}$ versus $_{uGs}$ stability map (air–water flow in an uncoated monolith column with d = 1.46 mm and a shower-head distributor) (fromKreutzer, 2003).

Kreutzer et al. (2005a) found that using a static mixer distributor resulted in less maldistribution than using a nozzle. The combination of a static mixer and a monolith is also applied in industry (Welp et al., 2002).Satterfield and Özel (1977) recommended using several layers of stacked monolith slices, rotated with respect to each other, to arrive at the best distribution. This slice-distributor might be conceived as a static mixer, as its working principle is the cutting and recombining of fluid flows without moving parts that is essentially the same as in a static mixer. The distributor as proposed by Satterfield and Özel (1977) was used by Kreutzer et al. (2001) to quantifiably improve the distribution in a pilot-reactor. So, several distributor designs are available in academic and patent literature alike. Although there is definitely room for improvement, the adverse effects of maldistribution can be overcome by the use of a good distributor: Peclet numbers of over a hundred are well possible for all but the shortest columns (Kreutzer et al., 2005afound a minimum of 2–3 m), and the gas-to-solid mass transfer is not affected by holdup distribution, as was demonstrated earlier.

SUGGESTION FOR FUTURE RE-SEARCH

In Table 1, the current state of the art is summarised, and references are listed that will assist in reactor design. With the current state-of-the-art, a reliable design of multiphase monolith reactors is possible. While some correlations need careful confirmation, or perhaps some experiments for different channel geometries or different fluids, the understanding of all relevant phenomena is robust enough for confidence.

Table 1: State-of-the-art in the reactor engineering of segmented flow in microchannels

Phenomenon	Status	References for design use
Film thickness (round channels)	Understood in detail, often hard-to-quantify Marangoni effects. Abundant experimental data	For clean liquid or Ca > 10^{-4}: Bretherton, 1961 and Aussilous and Quére, 2000; for real-life applications compare with e.g., Irandoust and Andersson, 1989a and Halpern et al., 1998
Film thickness (square channels)	Rough estimates available. Limited experimental data	Kolb and Cerro, 1991, Thulasidas et al., 1995a and Hazel and Heil, 2002
Pressure drop	Understood in detail, slug and bubble length must be known for design. Experimental data abundant, but usually without bubble and slug length	Kreutzer et al., 2003 and Kreutzer et al., 2005a
Gas–liquid mass transfer	CFD results provide mechanistic insight. Experimental data for liquids other than water still missing. Slug and bubble length dependence understood	Berčič and Pintar, 1997 and van Baten and Krishna, 2004
Liquid mass transfer to wall	CFD results provide mechanistic insight. Only one good source of experimental data	Kreutzer, 2003 and van Baten and Krishna, 2005, data from Horvath et al. (1973)
Gas mass transfer to wall	CFD results provide mechanistic insight. Most experiments under reacting conditions not fully mass transfer limited. Design uncertainty dominated by uncertainty in film thickness	Kreutzer, 2003 and Kreutzer et al., 2001
Residence time distribution (single channel)	Well understood, needs mass transfer correlation. Mildly slug length dependent	Thulasidas et al., 1995b and Thulasidas et al., 1999

Residence time distribution (multichannel)	Guidelines available from pressure drop model. Depends on distributor design and bubble and slug length	Kreutzer et al. (2005a)
Slug and bubble length	Complex dependence on entry geometry, must be determined for each setup. Easy to measure in single channel. Difficult in multichannel systems	Pressure drop seems feasible tool to measure-slug length: Kreutzer et al. (2003), Also conductivity:Heiszwolf et al. (2001b) and tomographic techniques:Mantle et al., 2002 and Gladden et al., 2003

For future research, the following subjects are suggested:

- For round channels, good estimates for the film thickness are available. For other channel cross-sections, e.g. square channels with rounded corners, accurate correlations for film thickness (and hence mass transfer) must still be formulated. With respect to mass transfer, the relevant transfer steps and their behaviour have been identified, but the lack of accurate film thickness predictions for non-circular channels makes the estimate of mass transfer coefficients problematic. Another approach to this problem is using round monolith channels, which results in more predictable and better mass transfer. Recent patent literature seems to indicate that this is the direction industry is taking (Heibel et al., 2003).

- The pressure drop model, combined with the stability analysis, provides several clues that will elucidate aspects of reactor scale-up. The amount of data in monolith columns is still limited, and the interaction between RTD and pressure drop must be worked out with greater detail experimentally.

Clearly, these suggested areas of research subjects will provide refined correlations for design purposes.

CONCLUSIONS

The hydrodynamics and transport phenomena relevant to designing co-current multiphase monoliths reactors and microreactors have been reviewed. The fundamental characteristics of the flow of elongated bubbles are now well understood: physically sound theory can be used to account for the vast majority of observations and very accurate predictions of the performance can often be made.

The motivation for the use of monoliths (or capillary channels in other configurations), and the proof of concept has been established with great certainty: an array of parallel straight channel provides a controlled reaction environment that combines high mass transfer, short intra-particle diffusional distances, extremely low pressure drop and minimal backmixing. These confirmed and understood observations make monoliths very promising packings for multiphase reactor columns, and predict a flourishing future for this flow pattern in labs-on-a-chip.

REFERENCES

1. Akbar, M.K., Plummer, D.A., Ghiaasiaan, S.M., 2003. On gas–liquid two-phase flow regimes in microchannels. International Journal of Multiphase Flow 29 (5), 855–865.

2. Aussilous, P., Quére, D., 2000. Quick deposition of a fluid on the wall of a tube. Physics of Fluids 12 (10), 2367–2371.

3. Berci͝ c, G., Pintar, A., 1997. The role of gas bubbles and liquid slug lengths ͝ on mass transport in the Taylor flow through capillaries. Chemical Engineering Science 52 (21,22), 3709–3719.

4. Bilek, A.M., Dee, K.C., Gaver III, D.P., 2003. Mechanisms of surfacetension-induced epithelial cell damage in a model of pulmonary airway reopening. Journal of Applied Physiology 94, 770–783.

5. Boger, T., Roy, S., Heibel, A.K., Borchers, O., 2003. A monolith loop reactor as an attractive alternative to slurry reactors. Catalysis Today 79–80, 441–451.

6. Boger, T., Heibel, A.K., Sorensen, C.M., 2004a. Monolithic catalysts for the chemical industry. Industrial and Engineering Chemistry Research 43 (16), 4602–4611.

7. Boger, T., Zieverink, M.M.P., Kreutzer, M.T., Kapteijn, F., Moulijn, J.A., Addiego, W.P., 2004b. Monolithic catalysts as an alternative to slurry systems: Hydrogenation of edible oil. Industrial and Engineering Chemistry Research 43 (16), 2337–2344.

8. Bos, C., Hoofd, L., Oostendorp, T., 1996. Reconsidering the effect of local plasma convection in a classical model of oxygen transport in capillaries. Microvascular Research 51 (3), 39–50.

9. Bretherton, F.P., 1961. The motion of long bubbles in tubes. Journal of Fluid Mechanics 10, 166–188.

10. Broekhuis, R.R., Machado, R.M., Nordquist, A.F., 2001. The ejectordriven monolith loop reactor—experiments and modelling. Catalysis Today 69 (1–4), 93–97.

11. Broekhuis, R.R., Budhlall, B.M., Nordquist, A.F., 2004. Monolith catalytic process for producing sorbitol: catalyst development and evaluation. Industrial and Engineering Chemistry Research 43 (17), 5146–5155.

12. Chang, H.-C., 2002. Bubble/drop transport in microchannels. In: The MEMS Handbook. CRC Press, Boca Raton. (Chapter 11). Chen, J.-D., 1986. Measuring the film thickness surrounding a bubble inside a capillary. Journal of Colloid and Interface Science 109 (2), 341–349.

13. Colburn, A.P., 1933. A method of correlating forced convection heat transfer data and a comparison with fluid friction. Transactions of the A.I.Ch.E. 29, 174–210.

14. Cox, B.G., 1963. On driving a viscous fluid out of a tube. Journal of Fluid Mechanics 14, 81–96.

15. Cox, B.G., 1964. An experimental investigation of the streamlines in viscous fluid expelled from a tube. Journal of Fluid Mechanics 20, 193–200.

16. Crezee, E., Barendregt, A., Kapteijn, F., Moulijn, J.A., 2001. Carbon coated monolithic catalysts in the selective oxidation of cyclohexanone. Catalysis Today 69, 283–290.

17. Cubaud, T., Ho, C.-M., 2004. Transport of bubbles in square microchannels. Physics of Fluids 16 (12), 4575–4585.

18. Cui, Z.F., Chang, S., Fane, A.G., 2003. The use of gas bubbling to enhance membrane processes—a review. Journal of Membrane Science 221, 1–35.

19. Cybulski, A., Stankiewicz, A., Edvinsson Albers, R.K., Moulijn, J.A., 1999. Monolithic reactors for fine chemicals industries: A comparative analysis of a monolithic reactor and a mechanically agitated slurry reactor. Chemical Engineering Science 54 (13–14), 2351–2358.

20. De Deugd, R.M., Chougule, R.B., Kreutzer, M.T., Meeuse, F.M., Grievink, J., Kapteijn, F., Moulijn, J.A., 2003. Is a monolithic loop reactor a viable option for Fischer-Tropsch synthesis? Chemical Engineering Science 58 (3–6), 583–591.

21. De Lathouder, K.M., Bakker, J.J.W., Kreutzer, M.T., Kapteijn, F., Moulijn, J.A., Wallin, S.A., 2004. Structured reactors for enzyme immobilization: advantages of tuning the wall morphology. Chemical Engineering Science 59 (22–23), 5027–5033.

22. Dowe, D.C., Rezkallah, K.S., 1999. Flow regime identification in microgravity two-phase flows using void fraction signals. International Journal of Multiphase Flow 25 (3), 433–457.

23. Duda, J.L., Vrentas, J.S., 1971a. Heat transfer in a cylindrical cavity. Journal of Fluid Mechanics 45, 261–279.

24. Duda, J.L., Vrentas, J.S., 1971b. Steady flow in the region of closed streamlines in a cylindrical cavity. Journal of Fluid Mechanics 45, 247–260.

25. Eberle, H.-J., Breimair, J., Domes, H., Gutermuth, T., 2000. Post reactor technology in phthalic anhydride plants. Petroleum Technology Quarterly (3), 129–133.

26. Edvinsson, R., 1994. Monolith reactors in three-phase processes. Ph.D. Thesis, Chalmers University of Technology, Goteborg, Sweden.

27. Edvinsson, R.K., Cybulski, A., 1995. A comparison between the monolithic reactor and the trickle-bed reactor for liquid phase hydrogenations. Catalysis Today 24, 173–179.

28. Edvinsson, R.K., Irandoust, S., 1993. Hydrodesulphurization of dibenzothiophene in a monolithic catalyst reactor. Industrial and Engineering Chemistry Research 32 (2), 391–395.

29. Edvinsson, R.K., Irandoust, S., 1996. Finite-element analysis of Taylor flow. A.I.Ch.E. Journal 42 (7), 1815–1823.

30. Edvinsson Albers, R., Nyström, M., Siverström, M., Sellin, A., Dellve, A.C., Andersson, U., Herrmann, W., Berglin, T., 2001. Development of a monolith-based process for H_2O_2 production: from idea to largescale implementation. Catalysis Today 69 (1–4), 247–252.

31. Fairbrother, F., Stubbs, A.E., 1935. The bubble-tube method of measurement. Journal of the Chemical Society 1, 527–529.

32. Fujioka, H., Grotberg, J.B., 2004. Steady propagation of a liquid plug in a two-dimensional channel. Journal of Biomechanical Engineering 126 (5), 567–577.

33. Fukano, T., Kariyasaki, A., 1993. Characteristics of gas–liquid two-phase flow in a capillary tube. Nuclear Engineering and Design 141, 59–68.

34. Galbiati, L., Andreini, P., 1992. Flow pattern transition for vertical downward two-phase flow in capillary tubes. Inlet mixing effects. International Communications in Heat and Mass Transfer 19 (6), 791–799.

35. Ghadiali, S.N., Gaver III, D.P., 2003. The influence of non-equilibrium surfactant dynamics on the flow of a semi-infinite bubble in a rigid cylindrical capillary tube. Journal of Fluid Mechanics 478, 165–169.

36.   Giavedoni, M.D., Saita, F.A., 1997. The axisymmetric and plane case of a gas phase steadily displacing a Newtonian liquid—a simultaneous solution to the governing equations. Physics of Fluids 9 (8), 2420– 2428.

37.   Giavedoni, M.D., Saita, F.A., 1999. The rear meniscus of a long bubble steadily displacing a Newtonian liquid in a capillary tube. Physics of Fluids 11 (4), 786–794.

38.   Gladden, L.F., 2004. Magnetic resonance: Ongoing and future role in chemical engineering research. A.I.Ch.E. Journal 49 (1), 2–9.

39.   Gladden, L.F., Lim, M.H.M., Mantle, M.D., Sederman, A.J., Stitt, E.H., 2003. MRI visualisation of two-phase flow in structured supports and trickle-bed reactors. Catalysis Today 79 (1–4), 203–210.

40.   Grolman, E., Edvinsson, R., Stankiewicz, A., Moulijn, J., 1996. Hydrodynamic instabilities in gas–liquid monolithic reactors. In: Proceedings of the ASME Heat Transfer Division, vol. 334-3. ASMEHTD, pp. 171–178.

41.   Grotberg, J.B., 2001. Respiratory fluid mechanics and transport processes. Annual Review of Biomedical Engineering 3, 421–457.

42.   Gruber, R., Melin, T., 2003. Radial mass-transfer enhancement in bubbletrain flow. International Journal of Heat and Mass Transfer 46 (15), 2799–2808.

43.   Günther, A., Khan, S.A., Thalmann, M., Trachsel, F., Jensen, K.F., 2004. Transport and reaction in microscale segmented gas liquid flow. Lab on a Chip 4, 278–286.

44.   Haakana, T., Kolehmainen, E., Turunen, I., Mikkola, J.-P., Salmi, T., 2004. The development of monolith reactors: general strategy with a case study. Chemical Engineering Science 59, 5629–5635.

45.   Halpern, D., Jensen, O.E., Grotberg, J.B., 1998. A theoretical study of surfactant and liquid delivery into the lung. Journal of Applied Physiology 85 (1), 333–352.

46. Hatziantoniou, V., Andersson, B., 1982. Solid-liquid mass transfer in segmented gas–liquid flow through a capillary. Industrial and Engineering Chemistry Fundamentals 21 (4), 451–456.

47. Hatziantoniou, V., Andersson, B., 1984. The segmented two phase flow monolithic catalyst reactor. An alternative for liquid phase hydrogenations. Industrial and Engineering Chemistry Fundamentals 23, 82–88.

48. Hatziantoniou, V., Andersson, B., Schöön, N.-H., 1986. Mass transfer and selectivity in liquid-phase hydrogenation of nitro compounds in a monolithic catalyst reactor with segmented gas–liquid flow. Industrial and Engineering Chemistry Process Design and Development 25 (4), 964–970.

49. Hazel, A.L., Heil, M., 2002. The steady propagation of a semi-infinite bubble into a tube of elliptical or rectangular cross-section. Journal of Fluid Mechanics 470, 91–114. M.T. Kreutzer et al. / Chemical Engineering Science 60 (2005) 5895 – 5916 5915

50. Heibel, A.K., Scheenen, T.W.J., Heiszwolf, J.J., van As, H., Kapteijn, F., Moulijn, J.A., 2001.

51. Gas and liquid phase distribution and their effect on reactor performance in the monolith film flow reactor. Chemical Engineering Science 56, 5935–5944.

52. Heibel, A.K., Kapteijn, F., Moulijn, J.A., 2002. Flooding performance of square channel monolith structures. Industrial and Engineering Chemistry Research 41 (26), 6759–6771.

53. Heibel, A.K., Liu, W., Morse, M., 2003. Structured catalysts and processes for gas–liquid reactions. World patent WO 03/041852 A1.

54. Heibel, A.K., Jamison, J.A., Woehl, P., Kapteijn, F., Moulijn, J.A., 2004a. Improving flooding performance for countercurrent monolith reactors. Industrial and Engineering Chemistry Research 43 (16), 4848–4855.

55. Heibel, A.K., Vergeldt, F.J., van As, H., Kapteijn, F., Moulijn, J.A., Boger, T., 2004b. Gas and liquid distribution in the

monolith film flow reactor. A.I.Ch.E. Journal 49 (12), 3007–3017.

56. Heil, M., 2001. Finite Reynolds number effects in the Bretherton problem. Physics of Fluids 13 (9), 2517–2521.

57. Heiszwolf, J.J., Engelvaart, L.B., van der Eijnden, M.G., Kreutzer, M.T., Kapteijn, F., Moulijn, J.A., 2001a. Hydrodynamic aspects of the monolith loop reactor. Chemical Engineering Science 56 (3), 805–812.

58. Heiszwolf, J.J., Kreutzer, M.T., van der Eijnden, M.G., Kapteijn, F., Moulijn, J.A., 2001b. Gas–liquid mass transfer of aqueous Taylor flow in monoliths. Catalysis Today 69 (1-4), 51–55.

59. Higbie, R., 1935. The rate of absorption of a pure gas into a still liquid during short periods of exposure. Transactions of the A.I.Ch.E. 31, 365–389.

60. Hoogendoorn, C.J., Lips, J., 1965. Axial mixing of liquid in gas–liquid flow through packed beds. Canadian Journal of Chemical Engineering 43, 125–131.

61. Horvath, C., Solomon, B.A., Engasser, H.-M., 1973. Measurement of radial transport in slug flow using enzyme tubes. Industrial and Engineering Chemistry Fundamentals 12 (4), 431–439.

62. Irandoust, S., 1989. The monolithic catalyst reactor. Ph.D. Thesis, Chalmers University of Technology, Göteborg, Sweden.

63. Irandoust, S., Andersson, B., 1988. Mass transfer and liquid–phase reactions in a segmented two-phase flow monolithic catalyst reactor. Chemical Engineering Science 43 (8), 1983–1988.

64. Irandoust, S., Andersson, B., 1989a. Liquid film in Taylor flow through a capillary. Industrial and Engineering Chemistry Research 28, 1684.

65. Irandoust, S., Andersson, B., 1989b. Simulation of flow and mass transfer in Taylor flow though a capillary. Computers and Chemical Engineering 13 (4/5), 519–526.

66. Irandoust, S., Andersson, B., Bengtsson, E., Silverstrom, M., 1989. Scale up of a monolytic catalyst reactor with two-phase flow. Industrial and Engineering Chemistry Research 28, 1489–1493.

67. Irandoust, S., Ertlé, S., Andersson, B., 1992. Gas liquid mass transfer of Taylor flow through a capillary. Canadian Journal of Chemical Engineering 70, 115.

68. Jayawardena, S.S., Balakotaiah, V., Witte, L., 1997. Pattern transition maps for microgravity two-phase flows. A.I.Ch.E. Journal 43 (6), 1637–1640.

69. Khan, S.A., Günther, A., Schmidt, M.A., Jensen, K.F., 2004. Microfluidic synthesis of colloidal silica. Langmuir 20, 8604–8611.

70. Klinghoffer, A.A., Cerro, R.L., Abraham, M.A., 1998. Influence of flow properties on the performance of the monolith froth reactor for catalytic wet oxidation of acetic acid. Industrial and Engineering Chemistry Research 37 (4), 1203–1210.

71. Kolb, W.B., Cerro, R.L., 1991. Coating the inside of a capillary of square cross-section. Chemical Engineering Science 46 (9), 2181–2195.

72. Kreutzer, M.T., 2003. Hydrodynamics of Taylor flow in capillaries and monoliths channels. Doctoral dissertation. Delft University of Technology, Delft, The Netherlands. Kreutzer, M.T., Du, P.,

73. Heiszwolf, J.J., Kapteijn, F., Moulijn, J.A., 2001. Mass transfer characteristics of three phase monolith reactors. Chemical Engineering Science 56 (22), 6015–6023.

74. Kreutzer, M.T.,

75. Heiszwolf, J.J., Kapteijn, F., Moulijn, J.A., 2003. Pressure drop of Taylor flow in capillaries: impact of slug length. In: Proceedings of the First International Conference on Microchannels and Minichannels. A.S.M.E, Rochester NY, U.S.A, pp. 153–159.

76. Kreutzer, M.T., Bakker, J.J.W., Kapteijn, F., Moulijn, J.A., Verheijen, P. J.T., 2005a. Scaling–up multiphase monolith

reactors: Linking residence time distribution and feed maldistribution using isobars. Industrial and Engineering Chemistry Research, DOI:10.1021/ie0492350, in press.

77. Kreutzer, M.T., Kapteijn, F., Moulijn, J.A., Kleijn, C.R., 2005b. Inertial and interfacial effects on pressure drop of Taylor flow in capillaries. A.I.Ch.E. Journal, in press.

78. Kreutzer, M.T., van der Eijnden, M.G., Kapteijn, F., Moulijn, J.A.,

79. Heiszwolf, J.J., 2005c. The pressure drop experiment to determine slug lengths in monoliths. Catalysis Today, accepted for publication.

80. Laborie, S., Cabassud, C., Durand-Bourlier, L., Lainé, J.M., 1998. Fouling control by air sparging inside hollow fibre membranes-effects on energy consumption. Desalination 118, 189–196.

81. Laborie, S., Cabassud, C., Durand-Bourlier, L., Lainé, J.M., 1999. Characterisation of gas–liquid two-phase flow inside capillaries. Chemical Engineering Science 54, 5723–5735.

82. Lebens, P.J., 1999. Development and design of a monolith reactor for gas–liquid countercurrent operation. Doctoral Dissertation, Delft University of Technology.

83. Liu, W., 2002. Ministructured catalyst bed for gas–liquid–solid multiphase catalytic reaction. A.I.Ch.E. Journal 48 (7), 1519–1532.

84. Machado, R.M., Parrillo, D.J., Boehme, R.P., Broekhuis, R.R., 1999. Use of a monolith catalyst for the hydrogenation of dinitrotoluene to toluenediamine. US Patent 6005143.

85. Mantle, M.D., Sederman, A.J., Gladden, L.F., 2002. Dynamic MRI visualization of two-phase flow in a ceramic monolith. A.I.Ch.E. Journal 48 (4), 909–912.

86. Marwan, H., Winterbottom, J.M., 2004. The selective hydrogenation of butyne-1,4-diol by supported palladiums: a comparative study on slurry, fixed bed, and monolith downflow bubble column reactors. Catalysis Today 97, 325–330.

87. Mazzarino, I., Baldi, G., 1987. Liquid phase hydrogenation on a monolith catalyst. In: Kulkarni, B., Mashelkar, R., Sharma, M. (Eds.), Recent Trends in Chemical Reaction Engineering. Wiley Eastern Ltd, New Delhi, p. 181.

88. Mishima, K., Hibiki, T., 1996. Some characteristics of air–water twophase flow in small diameter vertical tubes. International Journal of Multiphase Flow 22 (4), 703–712.

89. Natividad, R., Kulkarni, R., Nuithitikul, K., Raymahasay, S., Wood, J., Winterbottom, J.M., 2004. Analysis of the performance of single capillary and multiple capillary (monolith) reactors for the multiphase Pd-catalyzed hydrogenation of 2-butyne-1,4-diol. Chemical Engineering Science 59, 5431–5438.

90. Nijhuis, T.A., Beers, A.E.W., Vergunst, T., Hoek, I., Kapteijn, F., Moulijn, J.A., 2001a. Preparation of monolithic catalysts. Catalysis Reviews—Science and Engineering 43 (4), 345–380.

91. Nijhuis, T.A., Kreutzer, M.T., Romijn, A.C.J., Kapteijn, F., Moulijn, J.A., 2001b. Monolith catalysts as efficient three-phase reactors. Chemical Engineering Science 56 (3), 823–829.

92. Nijhuis, T.A., Dautzenberg, F.M., Moulijn, J.A., 2003a. Modelling of monolithic and trickle–bed reactors for the hydrogenation of styrene. Chemical Engineering Science 58, 1113–1124.

93. Nijhuis, T.A., van Koten, G., Moulijn, J.A., 2003b. Optimized palladium catalyst systems for the selective liquid-phase hydrogenation of functionalyzed alkynes. Applied Catalysis A: General 238 (2), 259–271.

94. Olbricht, W.L., 1996. Pore-scale prototypes of multiphase flow in porous media. Annual Review of Fluid Mechanics 28 (1), 187–213.

95. Oliver, D.R., Youngh-Hoon, A., 1968. Two-phase non-Newtonian flow—Part II: Heat transfer. Transactions of the Institution of Chemical Engineers 46, T116–T122.

96. Park, C.W., 1992. Influence of soluble surfactants on the motion of a finite bubble in a capillary tube. Physics of Fluids A 4 (11), 2335–2346. 5916 M.T. Kreutzer et al. / Chemical Engineering Science 60 (2005) 5895 – 5916

97. Pedersen, H., Horvath, C., 1981. Axial dispersion in a segmented gas–liquid flow. Industrial and Engineering Chemistry Fundamentals 20, 181–186.

98. Quan, X., Shi, H., Zhang, Y., Wang, J., Qian, Y., 2003. Biodegradation of 2,4-dichlorophenol in an air-lift honeycomb-like ceramic reactor. Process Biochemistry 38, 1545–1551.

99. Ratulowski, J., Chang, H.-C., 1989. Transport of bubbles in capillaries. Physics of Fluids A 1 (10), 1642–1655.

100. Ratulowski, J., Chang, H.-C., 1990. Marangoni effects of trace impurities on the motion of long gas bubbles in capillaries. Journal of Fluid Mechanics 210, 303–328.

101. Reinecke, N., Mewes, D., 1999. Oscillatory transient two-phase flows in single channels with reference to monolithic catalyst supports. International Journal of Multiphase Flow 25 (6–7), 1373–1393.

102. Reinelt, D.A., 1987. The rate at which a long bubble rises in a vertical tube. Journal of Fluid Mechanics 175, 557–565.

103. Reynolds, O., 1886. On the theory of lubrication and its application to Mr. Beauchamp tower's experiments, including an experimental determination of the viscosity of olive oil. Philosophical Transactions of the Royal Society of London 177, 190 sqq.

104. Rezkallah, K.S., 1996. Weber number based flow-pattern maps for liquidgas flows at microgravity. International Journal of Multiphase Flow 22 (6), 1265–1270.

105. Satterfield, C.N., Özel, F., 1977. Some characteristics of two-phase flow in monolithic catalyst structures. Industrial and Engineering Chemistry Fundamentals 16, 61–67.

106. Schutt, B.D., Serrano, B., Cerro, R.L., Abraham, M.A., 2002. Production of chemicals from cellulose and biomass-derived

compounds through catalytic sub-critical water oxidation in a monolith reactor. Biomass & Bioenergy 22, 365–375.

107. Severino, M., Giavedoni, M.D., Saita, F.A., 2003. A gas phase displacing a liquid with soluble surfactants out of a small conduit: the plane case. Physics of Fluids 15 (10), 2961–2972.

108. Shen, E.I., Udell, K.S., 1985. A finite element study of low Reynolds number two-phase flow in cylindrical tubes. Journal of Applied Mechanics 52 (1985), 253–256.

109. Smits, H.A., Stankiewicz, A., Glasz, W.C., Fogl, T.H.A., Moulijn, J.A., 1996. Selective three-phase hydrogenation of unsaturated hydrocarbons in a monolithic reactor. Chemical Engineering Science 51 (11), 3019–3025.

110. Song, H., Tice, J.D., Ismagilov, R.F., 2004. A microfluidic system for controlling reaction networks in time. Angewandte Chemie International Edition 42, 767–772.

111. Stebe, K.J., Barthes-Biesel, D., 1995. Marangoni effects of adsorption–desorption controlled surfactants on the leading end of an infinitely long bubble in a capillary. Journal of Fluid Mechanics 286, 25–48.

112. Suo, M., Griffith, P., 1964. Two phase flow in capillary tubes. Journal of Basic Engineering 86, 576–582.

113. Taha, T., Cui, Z., 2004. Hydrodynamics of slug flow inside capillaries. Chemical Engineering Science 59 (6), 1181–1190.

114. Taylor, G.I., 1961. Deposition of a viscous fluid on the wall of a tube. Journal of Fluid Mechanics 10, 161–165.

115. Thiers, R.E., Reed, A.H., Delander, K., 1971. Origin of the lag phase of continuous-flow analyis curves. Clinical Chemistry 17 (1), 42–48.

116. Thulasidas, T.C., Abraham, M.A., Cerro, R.L., 1995a. Bubble-train flow in capillaries of circular and square cross section. Chemical Engineering Science 50 (2), 183–199.

117. Thulasidas, T.C., Cerro, R.L., Abraham, M.A., 1995b. The monolith froth reactor: residence time modelling and analysis. Chemical Engineering Research and Design 73, 314–319.

118. Thulasidas, T.C., Abraham, M.A., Cerro, R.L., 1997. Flow patterns in liquid slugs during bubble-train flow inside capillaries. Chemical Engineering Science 52 (17), 2947–2962.

119. Thulasidas, T.C., Abraham, M.A., Cerro, R.L., 1999. Dispersion during bubble-train flow in capillaries. Chemical Engineering Science 54 (1), 61–76.

120. Triplett, K.A., Ghiaasiaan, S.M., Abdel-Khalik, S.I., Sadowski, D.L., 1999. Gas–liquid two-phase flow in microchannels, Part I: two-phase flow patterns. International Journal of Multiphase Flow 25, 377–394.

121. Valdés-Solís, T., Linders, M.J.G., Kapteijn, F., Marban, G.B.F.A., 2004. Adsorption and breakthrough performance of carbon-coated ceramic monoliths at low concentration of n-Butane. Chemical Engineering Science 59, 2791–2800.

122. Van Baten, J.M., Krishna, R., 2004. CFD simulations of mass transfer from Taylor bubbles rising in circular capillaries. Chemical Engineering Science 59, 2535–2545.

123. Van Baten, J.M., Krishna, R., 2005. CFD simulations of wall mass transfer for Taylor bubbles in circular capillaries. Chemical Engineering Science 60, 1117–1126.

124. Van Gulijk, C., Linders, M.J.G., Valés-Solís, T., Kapteijn, F., 2005. Intrinsic channel maldistribution in monolithic catalyst support structures. Chemical Engineering Journal, in press, doi:10.1016/j.cej.2005.03.013.

125. Vergunst, T., Linders, M., Kapteijn, F., Moulijn, J., 2001. Carbon-based monolithic structures. Catalysis Reviews—Science and Engineering 43 (3), 291–314.

126. Waters, S.L., Grotberg, J.B., 2002. The propagation of a surfactant laden liquid plug in a capillary tube. Physics of Fluids 14 (2), 471–480.

127. Welp, K.A., Cartolano, A.R., Parillo, D.J., Boehme, R.P., Machado, R.M., Caram, S., 2002. Monolith catalytic reactor coupled to static mixer. European Patent EP 1 287 884 A2.

128. Winterbottom, J.M., Marwan, H., Stitt, E.H., Natividad, R., 2003. The palladium catalysed hydrogenation of 2-butyne-1,4-diol in a monolith bubble column reactor. Catalysis Today 79–80, 391–399.

129. Wölk, G., Dreyer, M., Rath, H.J., 2000. Flow patterns in small diameter vertical non-circular channels. International Journal of Multiphase Flow 26 (6), 1037–1061.

130. Yawalkar, A.A., Sood, R., Kreutzer, M.T., Kapteijn, F., Moulijn, J.A., 2005. Axial mixing in monolith reactors: effect of channel size. Industrial and Engineering Chemistry Research 44 (7), 2046–2057.

131. Zhao, L., Rezkallah, K.S., 1993. Gas–liquid flow patterns at microgravity conditions. International Journal of Multiphase Flow 19 (5), 751–763.

Citations

CHAPTER 1

Bohuš Kysela, Ji í Konfršt, Ivan Fo t, and Zden k Chára, "CFD Simulation of the Discharge Flow from Standard Rushton Impeller,"International Journal of Chemical Engineering, vol. 2014, Article ID 706149, 7 pages, 2014. doi:10.1155/2014/706149.

CHAPTER 2

Nasim Salehi-Nik, Ghassem Amoabediny, Behdad Pouran, et al., "Engineering Parameters in Bioreactor's Design: A Critical Aspect in Tissue Engineering," BioMed Research International, vol. 2013, Article ID 762132, 15 pages, 2013. doi:10.1155/2013/762132.

CHAPTER 3

Chtourou, W. , Ammar, M. , Driss, Z. and Abid, M. (2014) CFD Prediction of the Turbulent Flow Generated in Stirred Square Tank by a Rushton Turbine. Energy and Power Engineering, 6, 95-110. doi: 10.4236/epe.2014.65010.

CHAPTER 4

A. Widiatmojo, K. Sasaki, N. Priagung Widodo and Y. Sugai, "Discrete Tracer Point Method to Evaluate Turbulent Diffusion in Circular Pipe Flow," Journal of Flow Control, Measurement & Visualization, Vol. 1 No. 2, 2013, pp. 57-68. doi: 10.4236/jfcmv.2013.12008.

CHAPTER 5

Grazia Monni, Mario De Salve, Bruno Panella, and Carlo Randaccio, "Electrical Capacitance Probe Characterization in Vertical Annular Two-Phase Flow," Science and Technology of Nuclear Installations, vol. 2013, Article ID 568287, 12 pages, 2013. doi:10.1155/2013/568287.

CHAPTER 6

S.L. Kiambi, H.K. Kiriamiti, A. Kumar, Characterization of two phase flows in chemical engineering reactors, Flow Measurement and Instrumentation, Volume 22, Issue 4, August 2011, Pages 265-271, ISSN 0955-5986, http://dx.doi.org/10.1016/j.flowmeasinst.2011.03.006.

CHAPTER 7

Ashish Karn, Christopher Ellis, Roger Arndt, Jiarong Hong, An integrative image measurement technique for dense bubbly flows with a wide size distribution, Chemical Engineering Science, Volume 122, 27 January 2015, Pages 240-249, ISSN 0009-2509, http://dx.doi.org/10.1016/j.ces.2014.09.036.

CHAPTER 8

Michiel T. Kreutzer, Freek Kapteijn, Jacob A. Moulijn, Johan J. Heiszwolf, Multiphase monolith reactors: Chemical reaction engineering of segmented flow in microchannels, Chemical Engineering Science, Volume 60, Issue 22, November 2005, Pages 5895-5916, ISSN 0009-2509, http://dx.doi.org/10.1016/j.ces.2005.03.022.

Index

A

Algebraic stress model (ASM) 71

C

Computational fluid dynamics
(CFD) 159
Computational Fluid Dynamics
(CFD) 67

D

Different technique 184
Discrete Tracer Point Method
(DTPM) 95, 115

E

Electrical capacitance probe (ECP)
123, 151
Experiments result 186

L

Large Eddy Simulation (LES) 1, 3
Laser Doppler Anemometry (LDA)
1, 7
Loss of coolant accidents (LOCAs)
124

M

Measurement method 14

Microchannels 233, 264, 266, 267, 268, 278, 283
Multiple reference frame (MRF) 71
Multi Reference Frame (MRF) 68

P

Particle Image Velocimetry (PIV) 189

Q

Quick-closing valves (QCVs) 130

R

Reaction engineering 215, 283
Reynolds Averaged Navier-Stokes (RANS) 2
Rushton turbines (RT) 68

S

Shadow Image Velocimetry technique (SIV) 188
Sliding Mesh (SM) 1, 3

T

Three-dimensional (3D) 20
Tissue engineering (TE) 20

Z

Zone of established flow (ZEF) 12
Zone of the established flow (ZEF) 13